Editorial

With the publication of Vol. 51 the editors and the publisher would like to take this opportunity to thank authors and readers for their collaboration and their efforts to meet the scientific requirements of this series. We appreciate the concern of our authors for the progress of "Advances in Polymer Science" and we also welcome the advice and critical comments of our readers.

With the publication of Vol. 51 we would also like to refer to a editorial policy: *this series publishes invited, critical review articles of new developments in all areas of polymer science in English (authors may naturally also include workes of their own)*. The responsible editor, that means the editor who has invited the author, discusses the scope of the review with the author on the basis of a tentative outline which the author is asked to provide. The author and editor are responsible for the scientific quality of the contribution.

Manuscripts must be submitted in content, language, and form satisfactory to Springer-Verlag. Figures and formulas should be reproducible. To meet the convenience of our readers, the publisher will include "volume index" which characterizes the content of the volume.

The editors and the publisher will make all efforts to publish the manuscripts as rapidly as possible, i.e., at the maximum six months after the submission of an accepted paper. Contributions from diverse areas of polymer science must occasionally be united in one volume. In such cases a "volume index" cannot meet all expectations, but will nevertheless provide more information than a mere volume number.

Starting with Vol. 51, each volume will contain a subject index.

Editors   Publisher

75   Advances in Polymer Science

Epoxy Resins and Composites II

Editor: K. Dušek

With Contributions by
J. P. Bell, L. T. Drzal, J. L. Koenig, E. Mertzel,
B. A. Rozenberg, R. G. Schmidt

With 78 Figures and 22 Tables

Springer-Verlag
Berlin Heidelberg New York Tokyo

ISBN-3-540-15825-1 Springer-Verlag Berlin Heidelberg New York Tokyo
ISBN-0-387-15825-1 Springer-Verlag New York Heidelberg Berlin Tokyo

Library of Congress Catalog Card Number 61-642

This work is subject to copyright. All rights are reserved, whether the whole or part of the material is concerned, specifically those of translation, reprinting, re-use of illustrations, broadcasting, reproduction by photocopying machine or similar means, and storage in data banks. Under § 54 of the German Copyright Law copies are made for other than private use, a fee is payable to "Verwertungsgesellschaft Wort", Munich.

© by Springer-Verlag Berlin · Heidelberg 1986
Printed in GDR

The use of registered names, trademarks, etc. in this publication does not imply, even in the absence of a specific statement, that such names are exempt from the relevant protective laws and regulations and therefore free for general use.

Typesetting and Offsetprinting: Th. Müntzer, GDR;
Bookbinding: Lüderitz & Bauer, Berlin
2154/3020-543210

Editors

Prof. Henri Benoit, CNRS, Centre de Recherches sur les Macromolecules, 6, rue Boussingault, 67083 Strasbourg Cedex, France

Prof. Hans-Joachim Cantow, Institut für Makromolekulare Chemie der Universität, Stefan-Meier-Str. 31, 7800 Freiburg i. Br., FRG

Prof. Gino Dall'Asta, Via Pusiano 30, 20137 Milano, Italy

Prof. Karel Dušek, Institute of Macromolecular Chemistry, Czechoslovak Academy of Sciences, 16206 Prague 616, ČSSR

Prof. John D. Ferry, Department of Chemistry, The University of Wisconsin, Madison, Wisconsin 53706, U.S.A.

Prof. Hiroshi Fujita, Department of Macromolecular Science, Osaka University, Toyonaka, Osaka, Japan

Prof. Manfred Gordon, Department of Pure Mathematics and Mathematical Statistics, University of Cambridge CB2 1SB, England

Prof. Gisela Henrici-Olivé, Chemical Department, University of California, San Diego, La Jolla, CA 92037, U.S.A.

Prof. Dr. habil. Günter Heublein, Sektion Chemie, Friedrich-Schiller-Universität, Humboldtstraße 10, 69 Jena, DDR

Prof. Dr. Hartwig Höcker, Universität Bayreuth, Makromolekulare Chemie I, Universitätsstr. 30, 8580 Bayreuth, FRG

Prof. Hans-Henning Kausch, Laboratoire de Polymères, Ecole Polytechnique Fédérale de Lausanne, 32, ch. de Bellerive, 1007 Lausanne, Switzerland

Prof. Joseph P. Kennedy, Institute of Polymer Science, The University of Akron, Akron, Ohio 44325, U.S.A.

Prof. Anthony Ledwith, Department of Inorganic, Physical and Industrial Chemistry, University of Liverpool, Liverpool L69 3BX, England

Prof. Seizo Okamura, No. 24, Minamigoshi-Machi Okazaki, Sakyo-Ku. Kyoto 606, Japan

Professor Salvador Olivé, Chemical Department, University of California, San Diego, La Jolla, CA 92037, U.S.A.

Prof. Charles G. Overberger, Department of Chemistry. The University of Michigan, Ann Arbor, Michigan 48 104, U.S.A.

Prof. Helmut Ringsdorf, Institut für Organische Chemie, Johannes-Gutenberg-Universität, J.-J.-Becher Weg 18-20, 6500 Mainz, FRG

Prof. Takeo Saegusa, Department of Synthetic Chemistry, Faculty of Engineering, Kyoto University, Kyoto, Japan

Prof. Günter Victor Schulz, Institut für Physikalische Chemie der Universität, 6500 Mainz, FRG

Prof. William P. Slichter, Chemical Physics Research Department, Bell Telephone Laboratories, Murray Hill, New Jersey 07971, U.S.A.

Prof. John K. Stille, Department of Chemistry. Colorado State University, Fort Collins, Colorado 80523, U.S.A.

Preface

This volume of ADVANCES IN POLYMER SCIENCE contains the first part of a series of critical reviews on selected topics concerning epoxy resins and composites. The last decade has been marked by an intense development of applications of epoxy resins in traditional and newly developing areas such as coatings, adhesives, civil engineering or electronics and high-performance composites. The growing interest in applications and requirements of high quality and performance has provoked a new wave in fundamental research in the area of resin synthesis, curing systems, properties of cured products and methods of their characterization.

The collection of reviews to be published in ADVANCES IN POLYMER SCIENCE is devoted just to these fundamental problems. The epoxy resin-curing agent formulations are typical thermosetting systems of a rather high degree of complexity. Therefore, some of the formation-structure-properties relationships are still of empirical or semiempirical nature. The main objective of this series of articles is to demonstrate the progress in research towards the understanding of these relationships in terms of current theories of macromolecular systems.

Because of the complexity of the problems discussed, the theoretical approaches and interpretation of results presented by various authors and schools may be somewhat different. It may be hoped, however, that a confrontation of ideas may positively contribute to the knowledge about this important class of polymeric materials.

In view of the wide range of this area, it was impossible to publish all contributions in successive volumes of ADVANCES IN POLYMER SCIENCE. Part I was published in Vol. 72; Part II in this Vol. 75. Part III and Part IV will follow in the beginning of 1986.

The reader may appreciate receiving a list of all contributions to the series EPOXY RESINS AND COMPOSITES to appear in ADVANCES IN POLYMER SCIENCE:

M. T. Aronhime and J. K. Gillham (Princeton University, Princeton, N.J., USA)

The Time-Temperature-Transformation (TTT) Cure Diagram of Thermosetting Polymeric Systems

A. Apicella and L. Nicolais (University of Naples, Naples, Italy) Effect of Water on the Properties of Epoxy Matrix and Composites (Part I, Vol. 72)

J. M. Barton (Royal Aircraft Establishment, Farnborough, UK): The Application of Differential Scanning Calorimetry (DSC) to the Study of Epoxy Resins Curing Reactions (Part I, Vol. 72)

W. Burchard (University of Freiburg, Freiburg i. Br., FRG) Branching in Epoxy Resins Based on Diglycidyl Ethers of Bisphenol A

L. T. Drzal (Michigan State University, East Lansing, MI, USA) The Interphase in Epoxy Composites (Part II, Vol. 75)

K. Dušek (Institute of Macromolecular Chemistry, Czechoslovak Academy of Sciences, Prague, Czechoslovakia) Network Formation in Curing of Epoxy Resins

M. Fedtke (Technical University, Merseburg, GDR) Elucidation of the Mechanism of Epoxy Curing by Model Reactions

A. Gupta (Jet Propulsion Laboratory, Caltech, Pasadena, CA, USA) Mechanism and Kinetics of the Cure Process in Tetraglycidylmethane Dianiline-Diaminodiphenyl Sulphone Thermoset System

T. Kamon and H. Furukawa (The Kyoto Municipal Research Institute of Industry, Kyoto, Japan) Curing Mechanism and Mechanical Properties of Cured Epoxy Resins

J. L. Kardos and M. P. Duduković (Washington University, St. Louis. MO, USA) Void Growth and Transport During Processing of Thermosetting Matrix Composites

A. J. Kinloch (Imperial College, London, UK) Mechanics and Mechanism of Fracture of Thermosetting Epoxy Polymers

E. S. W. Kong (Hewlett-Packard Laboratories, Palo Alto, CA, USA) Physical Aging in Epoxy Matrices and Composites

J. D. LeMay and F. N. Kelley (University of Akron, Akron, OH, USA) Structure and Untimate Properties of Epoxy Resisns

F. Lohse, and H. Zweifel (Ciba-Geigy, Basle, Switzerland) Photocrosslinking of Epoxy Resins

J. A. Manson, R. W. Hertzberg, G. Attalla, D. Shah, J. Hwang and J. Turkanis (Lehigh University, Bethlehem, PA, USA)
Fatigue in Neat and Rubber-Modified Epoxies

E. Mertzel and J. L. Koenig (Case Western Reserve University, Cleveland, OH, USA)
Application of FT-IR and NMR to Epoxy Resins (Part II, Vol. 75)

R. J. Morgan (Lawrence Livermore National Laboratory, Livermore, CA, USA)
Structure-Properties Relations of Epoxies Used as Composite Matrices (Part I, Vol. 72)

E. F. Oleinik (Institute of Chemical Physics, Academy of Sciences of USSR, Moscow, USSR)
Structure and Properties of Epoxy-Aromatic Amine Networks in the Glassy State

B. A. Rozenberg (Institute of Chemical Physics, Academy of Sciences of USSR, Moscow, USSR)
Kinetics, Thermodynamics and Mechanism of Reactions of Epoxy Oligomers with Amines (Part II, Vol. 75)

S. D. Senturia and N. F. Sheppard (Massachusetts Institute of Technology, Cambridge, MA, USA)
Dielectric Analysis of Epoxy Cure

R. G. Schmidt and J. P. Bell (University of Connecticut, Storrs, CT, USA)
Epoxy Adhesion to Metals (Part II, Vol. 75)

E. M. Yorkgitis, N. S. Eiss, Jr., C. Tran, G. L. Wilkes and J. E. McGrath (Virginia Polytechnic Institute, Blacksburg, VA, USA)
Siloxane Modified Epoxy Resins (Part I, Vol. 72)

The editor wishes to express his gratitude to all contributors for their cooperation.

Prague, November 1985
Karel Dušek
(Editor)

Table of Contents

The Interphase in Epoxy Composites
L. T. Drzal 1

Epoxy Adhesion to Metals
R. G. Schmidt, J. P. Bell. 33

Application of FT-IR and NMR to Epoxy Resins
E. Mertzel, J. L. Koenig 73

Kinetics, Thermodynamics and Mechanism of Reactions of Epoxy Oligomers with Amines
B. A. Rozenberg 113

Author Index Volumes 1–75 167

Subject Index 177

The Interphase in Epoxy Composites

Lawrence T. Drzal
Department of Chemical Engineering,
Center for Composite Materials and Structures,
Michigan State University,
East Lansing, Michigan 48824-1226/U.S.A.

This article introduces the concept of an interphase in epoxy composites as being an identifiable entity which in some circumstances can control composite properties. The review is divided into two major parts. The first describes the interphase from a phenomenological viewpoint as to its possible structure and composition. Examples are given which serve to illustrate the potential for inducing compositional and structural differences at the reinforcement surface-epoxy interface. The second portion uses the results of micromechanical modeling of ideal composites to illustrate the magnitude and type of forces which operate in the interphase region. Specific examples are included which quantify the relationship between interphase structure and composite mechanical and fracture properties.

1 Introduction 3

2 Interphase Constituents 4
   2.1 Epoxy Resin 4
   2.2 Crosslinking Agents 5
   2.3 Cured Epoxy Properties 6
   2.4 Reinforcement Surface 8
      2.4.1 Surface Structure and Chemistry 8
      2.4.2 Physisorbed Material 10
      2.4.3 Topography 13
      2.4.4 Adsorption of Epoxy Components 14
   2.5 Interfacial Additives 15
      2.5.1 Coupling Agents 15
      2.5.2 Finishes and Primers 15
   2.6 Wettability 16

3 Interphase Effects on Composite Strength 16
   3.1 Tensile and Compressive Strength 17
   3.2 Transverse Strength 19
   3.3 Interfacial Shear Strength 20

4 Interphase Effect on Composite Fracture 23
   4.1 Fiber-Matrix Debonding 23

4.2 Fiber Deformation and Fracture 24
  4.3 Fiber Pull-Out 24

5 Interphase Effects on Composite Environmental Resistance 27

6 Conclusions 30

7 References 30

1 Introduction

Over the last decade advances have occurred very rapidly in the area identified as composite materials. In general, a composite material is the combination of any two or more materials, one of which has superior mechanical properties but is in a difficult to use form (e.g. fiber, powder, etc.). The superior component is usually the reinforcement while the other component serves as the matrix in which the reinforcement is dispersed. The resultant composite is a material whose properties are near those of the reinforcement element but in a form which can be easily handled and can easily function as a structural element. Included in this definition are all of the reinforced materials including particulate, fiber, flake and sheet reinforcements. Adhesive joints for, example, would be a planar or two dimensional composite [1].

One of the largest class of composite materials in use is the continuous reinforced fiber composite using epoxy as the matrix material. Reinforcing elements can be glass, graphite or aramid fibers or for the adhesive joint aluminum, steel or titanium. In any of these composite applications, the coupling of the reinforcement with the surrounding epoxy matrix is a necessary element. Optimum bonding is responsible for maximum static and dynamic mechanical properties and environmental resistance. Indeed, interfacial adhesion between fiber and matrix is optimized for most commercial composites marketed today. However, because of the lack of a true understanding of the molecular mechanisms occurring at the fiber-matrix interface, new material combinations or substitutions for existing reinforcing fibers, adherends or epoxy matrices are difficult to achieve. This lack of ability to 'a priori' select surface and interfacial conditions is due to the lack of understanding of the structure and composition of the fiber-matrix interface itself.

In optimized commercial materials, the interface functions as an efficient transmitter of forces between fiber and matrix. As such, as long as the interface is intact, composite material behavior can be adequately described by models which assume ideal adhesion between fiber and matrix and consider the interface to be a two-dimensional boundary.

Recent work has expanded the concept of the fiber-matrix interface which exists as a two-dimensional boundary into that of a fiber-matrix interphase that exists in three dimensions [2]. The complexity of this interphase can best be illustrated with the use of a schematic model which allows the many different characteristics of this region to be enumerated as shown in Fig. 1 [3].

By definition, the interphase exists from some point in the fiber where the local properties begin to change from the fiber bulk properties, through the actual interface into the matrix where the local properties again equal the bulk properties. Within this region, various components of known and unknown effect on the interphase can be identified. The fiber may have morphological variations near the fiber surface which are not present in the bulk of the fiber. The surface area of the fiber can be much greater that its geometrical value because of pores or cracks present on the surface. The atomic and molecular composition of the fiber surface can be quite different from the bulk of the fiber. Surface treatments can add surface chemical groups or remove the original surface giving rise to a chemically and structurally different region. Exposure to air before composite processing can result in the adsorption of chemical species which may alter or eliminate certain beneficial surface reactivity. These adsorbed materials may also desorb at the elevated temperatures seen in composite fabrication

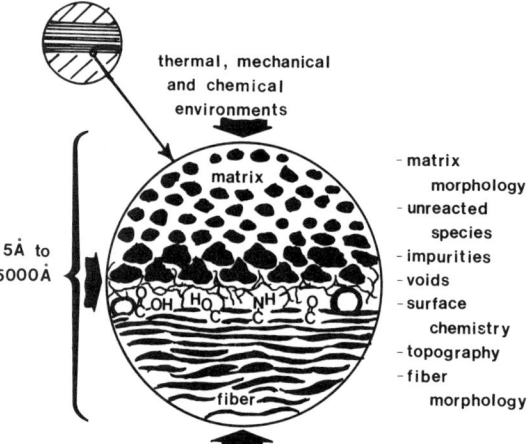

Fig. 1. A schematic model of the epoxy-reinforcement interphase highlighting the possible components, size and imposed environments acting in this region. From Drzal et al. [75]

and be a source of volatiles which disrupt the interface if not removed. Thermodynamic wetting of the fiber surface by the matrix is a necessary condition for fiber-matrix adhesion and is determined by the free energies of the components. Both chemical and physical bonds exist at the interface and the number and type of each strongly influences the interaction between fiber and matrix. The structure of the matrix in the interphase can be influenced by the proximity of the fiber surface. Changes in reactivity due to adsorption of matrix components can alter the local morphology. Unreacted matrix components and impurities can diffuse to the interphase region altering the local structure.

Each of these phenomena can vary in magnitude and can occur simultaneously in the interphase region. Depending on the material system the interphase itself can be composed of any or all of these components and can extend in thickness from a few to a few thousand nanometers. The structure of this region can have profound effects on the performance of the composite in terms of its mechanical strength, chemical and thermal durability. The exact nature of this region must be understood if accurate predictive models of interphase behavior are to be developed and integrated into a model of composite performance. Only when the exact nature of this region is understood well the interphase be considered as a composite variable which itself can be altered in a rational manner to optimize composite performance.

2 Interphase Constituents

2.1 Epoxy Resin

The epoxy resin can be defined as any molecule that contains two or more alpha-epoxy groups which can be reacted to form a thermoset system. An example of a difunctional epoxy resin is diglycidyl ether of Bisphenol-A (DGEBA) which is formed by the reaction of Bisphenol-A with epichlorohydrin in the presence of sodium hydroxide. The base catalyzes the reaction to produce the chlorohydrin intermediate, acts as the dehydrohalogenating agent, and nuetralizes the hydrochloric acid formed in the reaction [4]. Varying the ratio of reactants produces different relative amounts of the high molecular weight products in addition to the monomeric unit.

In addition to the Bisphenol-A backbone epoxy resins, epoxies with substituted aromatic backbones and in the tri- and tetra- functional forms have been produced. Structure-property relationships exist so that an epoxy backbone chemistry can be selected for the desired end product property. Properties such as oxygen permeability, moisture vapor transmission and glass transition temperature have been related to the backbone structure of epoxy resins [5]. Whatever the backbone structure, resins containing only the pure monomeric form can be produced but usually a mixture of different molecular weight species are present with their distribution being dictated by the end-use of the resin.

Considering both the chemistry of epoxy resins and the reactants required to produce it, impurities, unreacted raw materials, catalyst and other monomeric species can be expected to be present in the resin in low levels (i.e. < 100 ppm). Under the mechanical, thermal and chemical stress levels acting at the interface both during curing and after curing, concentration of these species at the interface could be quite high giving rise to an interface structure unrelated to the bulk.

2.2 Crosslinking Agents

The great utility of epoxy resins lies in the fact that the epoxide group is reactive with a large number of molecules which form tough thermoset networks without the evolution of by-products. The reactants can be classified as either direct participants in the crosslinked network or those which promote crosslinking catalytically.

The reactants which participate directly in the reaction and are incorporated in the crosslinked network can be further subdivided into basic or acidic curing agents. The basic curing agents include Lewis bases, inorganic bases, primary and secondary amines and amides. The acidic curing agents can be carboxylic acid anhydrides, polybasic organic acids, phenols and Lewis acids. These curing agents react by opening the epoxy ring either anionically or cationically.

The catalytically functioning curing agents do not directly participate in the crosslinked network but promote reactions between epoxy groups themselves. Tertiary amines as well as boron trifluoride type complexes are effective catalytic agents. Excellent discussions of the specific curing agents, their reactivity with epoxy and their effect on epoxy mechanical properties are available in the literature [4, 6].

In addition to the large choice of curing agents, the physical properties of the curing agents can affect and be affected by the processing of the epoxy system. The vapor pressure of curing agents may differ by a factor of ten at the same temperature [7]. The more volatile component can be removed during the vacuum assisted debulking step in matrix processing resulting in a material with less than the proper amount of crosslinking. The solubility of the various components can be quite different [8]. For example, dicyanamide, a solid curing agent has a low solubility in epoxy resins. Undissolved dicyanamide is ineffective as a curing agent, can act as a stress concentration point in the matrix and as a sink for water absorption because of its hydrophilic nature. The dicyanamide particles may also be selectively confined to certain regions of the composite because of their size.

Crosslinking agent purity can cause variations in both properties as well as local structure. Meta-phenylenediamine is a solid which readily dissolves in epoxy resin at low temperature. The amine is very sensitive to oxidation however, and can discolor with exposure to atmospheric oxygen, water and/or light. While not affecting the bulk matrix properties, the presence of this degraded curing agent has been shown to cause a reduced level of adhesion to glass fibers [9].

It is clear that other components quite different chemically from the main constituents of the epoxy resin system may be present in the starting material. The structure of the cured epoxy may or may not incorporate these components. To the extent that these other species are not part of the crosslinked epoxy network, they can be concentrated at the interphase or they may be able to migrate to the interphase during the curing process.

2.3 Cured Epoxy Properties

The matrix side of the interphase region has a structure that is the result of the interaction of an epoxy resin or resins and one or more of the different curing agents under the temperature and pressure conditions imposed by the processing cycle. Although a great deal is known about the specific interactions between the epoxy group and the various curing agents, little is known about the actual crosslink environment in an actual cured matrix. This is because various other factors may override specific molecular interactions when an epoxy material is processed. Commercial epoxies, for example, are formulated using combinations of resins as well as curing agents. It is not uncommon to have two or three epoxy resins combined with a curing agent blend of primary amines, inorganic acid complexes and latent catalytic curing agents. These formulations are developed to maximize mechanical properties and allow some flexibility in composite processing cycles. Because the components of the formulation differ in chemical and physical properties, processing conditions may alter the final composition and properties and can influence the matrix morphology and matrix mechanical properties.

Since volatility of the curing agent can cause changes in the final curing agent concentration, the possibility exists that the local crosslinking agent concentration may be different from the nominal concentration at the start of the processing procedure. A few studies have investigated the alteration of epoxy properties as a function of curing agent concentration. Selby and Miller [10] have studied the variation of fracture and mechanical properties of an epoxy cured with various amounts of diaminodiphenylmethane. They found that the fracture surface energy was a maximum at a value greater than the stoichiometric amount but that the tensile modulus, compressive modulus, and compressive yield stress were a minimum at stoichiometry. Kim et al [11] investigated the structure-property relationships in a methylene dianiline cured epoxy system. They also found a slight change in tensile strength, modulus and elongation with composition and a larger effect on the impact strengths and fracture toughness. The cured epoxy properties tended to have relative maxima or minima near the stoichiometric point. A characteristic heterogeneity in the structure of the cured systems was observed to vary in size with composition. Gupta et al. [12] have studied

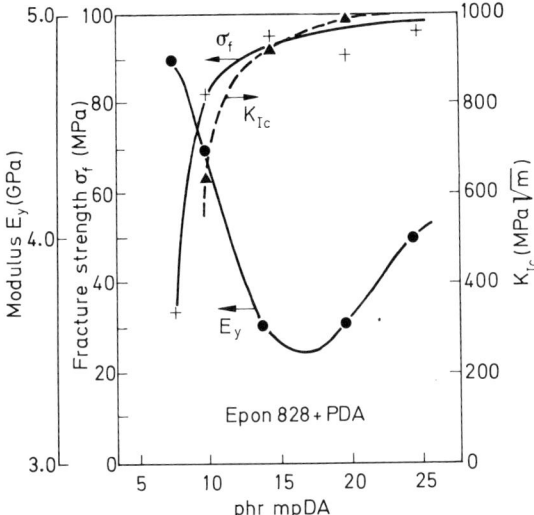

Fig. 2. Initial Tensile Modulus (E_y), Fracture Strength (σ_f), and Fracture Toughness (K_{Ic}) as a function of the curing agent concentration. From Drzal et al. [78]

the difuctional epoxy/meta-phenylene diamine system and have obtained similar results. The stoichiometric point corresponds to a maximum or minimum in various cured epoxy properties. Figure 2 is a plot of the initial tensile modulus, strain to failure and fracture toughness of this system as a function of curing agent concentration. While the strain to failure and the fracture toughness reach a relative maximum at the stoichiometric point of 14.5 parts per hundred (phr), the modulus reaches a minimum. Network heterogeneity has been observed for this system as a function of curing agent concentration. Figure 3 shows the difference in texture of the epoxy as a fuction of the curing agent concentration. The identifiable features range in size from 5 to 50 nanometers and can be detected even when different types of methods are used to prepare the cured epoxy surfaces.

The catalyzed system are much more complex both in terms of the reactions occurring in the system as well as the in the final structure. Byrne, Hagnauer and Schneider [13] have completed a detailed study of the interactions between components. A dependency of either epoxy-amine reaction or homopolymerization on both the catalyst concentration and processing conditions is reported. Morgan [14], Geil [15], Mijovic [16, 17], Wake [18] and others have studied similar systems and shown variations in heterogeneity with composition. Swetlin [19] has done extensive studies of epoxy structure and its relation to properties below the glass transition temperature as a function of curing agent concentration but has not detected or used heterogeneity to explain the variation in the cured epoxy properties with amine to epoxy ratio. Bell [20] has indicated that in some circumstances the presence of the heterogeneity can be eliminated by additional mixing. The issue of the existence of a heterogeneous feature which varies in average size with curing agent concentration and whether these heterogeneities are responsible for the variation in properties with concentration is not resolved and is under active investigation.

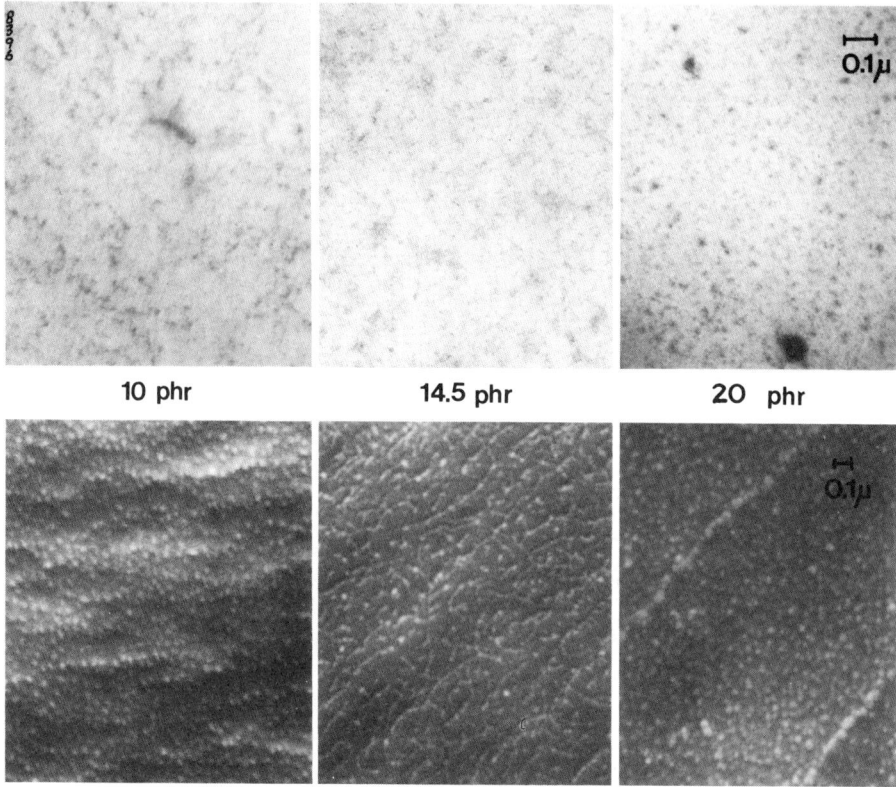

Fig. 3. Epoxy heterogeneities as a function of amine curing agent content determined by two different methods. The upper series were microtomed samples stained with osmium tetroxide and the lower series were plasma treated fracture surfaces. Both methods gave size and distribution values for the heterogeneities which agreed qualitatively

2.4 Reinforcement Surfaces

The value of the epoxy resins lies in their reactivity with a variety of chemical groups. This enhanced reactivity also means that the surface chemistry of the reinforcement which the epoxies are cured against, can alter the local structure in the interphase region [21]. The most common reinforcement surfaces cured in contact with the epoxies are carbon/graphite fibers, glass fibers, aramid fibers and metal oxides. The surface chemistry of these reinforcement surfaces is quite diverse and in many cases can be the reason for alteration of the interphase epoxy structure as compared to the bulk.

2.4.1 Surface Structure and Chemistry

Carbon/graphite fibers are prepared from either a polyacrylonitrile or rayon precursor fiber or from a pitch precursor [22, 23]. In either case, the fibers are treated at high temperature in an inert environment to produce a reinforcement fiber having a graphitic structure, the perfection of which determines the mechanical properties of the fiber. Although the fiber is almost entirely carbon, its surface may be chemically quite distinct from typical carbon surfaces. The carbon resides in turbostratic graphitic layers which are part of the graphitic crystallites which in turn are the basic unit of the fibers. These crytstallites are formed into ribbons which are oriented roughly parallel to the fiber axis. The width, thickness and undulating nature of these ribbons varies with the fiber heat treatment temperature [23, 24]. The orientation of these ribbons also varies with position in the fiber. Therefore, one would expect not only the inert graphitic basal plane to exist at the fiber surface, but the highly reactive corners and edges of these crystallites also. Because of the variation in fibril orientation with position, the surface of the fiber would be expected to be morphologically different from the bulk of the fiber. The variation with radial position would be most pronounced for low heat treatment and therefore low modulus fibers and less for the higher modulus fibers [25, 26]. The proportion of the surface composed of edge and corner areas would change with fiber modulus. This would then limit the maximum number of chemically different species that could be added to the fiber surface since the edges and corners are preferred sites for oxidative attack.

The quantity and type of chemical group present on the fiber surface is a function of the type of fiber, i.e. its graphitization temperature, and the type of surface treatment used, i.e. gaseous, liquid or catalytic oxidation [27]. Boehm et al. [28] have attempted to identify the type of chemical group present on the carbon surface. Using wet chemical techniques, they have shown that the oxygen present on the fiber surface can be in as many as four different chemical groups. Carboxylic acid, lactone, carbonyl and alcoholic oxygens are present initially or can be added to the surface through surface treatments to enhance adhesion to epoxies.

Recent surface spectroscopic work has shown that additional species besides oxygen are present on the fiber surface (Fig. 4). Nitrogen in the form of amine or cyano groups is almost always on the low heat treatment temperature fiber surface [29, 30]. Trace amounts or elements such as silicon, iron [31] can be present. The exact chemical nature of these groups is a subject still under investigation. Ishitami [32] and Proctor and Sherwood [33] have attempted to assign molecular environments to the oxygen and nitrogen groups. They have postulated the existence of nitrogen as amines. Drzal has identified sodium, trapped in the lower modulus fibers from the earlier polymer fiber spinning steps, as being present on the fiber surface and has shown that the sodium has the

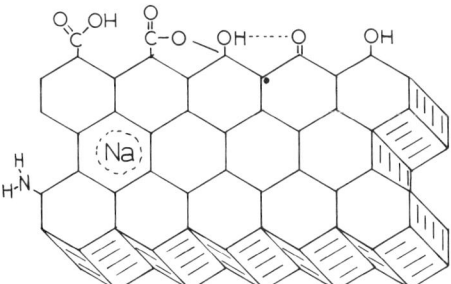

Fig. 4. Schematic illustration of the potential surface chemical groups which have been found on the surface of carbon fibers

ability to diffuse to the fiber surface from the bulk of the fiber under moderate elevated temperature conditions [34].

Glass fibers have a longer history as a reinforcement than carbon fibers. Their surface chemistry however, is more complex and probably less understood. Rynd and Rastogi [35] have used Auger surface spectroscopy to compare the 'E' and 'S' glass surface composition to that of the bulk. Several differences were noted in composition. For example, compared to the bulk concentration of 'E' glass fibers, the surface had an excess of fluorine, silicon and aluminum but was deficient in magnesium, boron and calcium. The 'S' glass fiber surface was rich in magnesium and aluminium. The surface atoms of glass are not in the free state but are chemically combined with their respective oxides.

The aramid fibers are polymeric reinforcing fibers made from poly(p-phenylene terephthalamide) that are used with epoxies to form structural composites. The fiber surfaces, composed of 500 to 700 nm fibrils [36], are relatively inert because the surface atoms are covalently bonded into the polymeric structure. Allred [37] has studied the surface of these fibers spectroscopically. He has found that the surface appears to be similar to that of an oxidized hydrocarbon. Nitrogen which makes up a significant part of the bulk composition of the polymer is in much lower concentration on the surface. The oxidized hydrocarbon surface may be due to additives present in the fiber from the processing steps which can now diffuse to the surface (sodium sulfate, stearic and palmitic acids [38]) or may be produced during the elevated temperature processing of the fiber. The layer consisting of this different material has been estimated to be about 2.5 nm in thickness. The surface chemistry of these aramid fibers can be altered however. Allred has shown that a plasma can be used to aminate the surface for example.

Metals such as aluminium, steel, and titanium are the primary adherends used for adhesively bonded structure. They are never bonded directly to a polymeric adhesive, however. A protective oxide, either naturally occurring or created on the metal surface either through a chemical etching or anodization technique is provided for corrosion protection. The resultant oxide has a morphology distinct from the bulk and a surface chemistry dependent on the conditions used to form the oxide [39]. Studies on various aluminum alloy compositions show that while the oxide composition is invariant with bulk composition, the oxide surface contains chemical species that are characteristic of the base alloy and the anodization bath [40-42].

Overall the surface chemical composition of the reinforcing fibers or the adherends is chemically quite different from the bulk composition of these materials. Specific interactions between epoxies and these surfaces without cognizance of the different surface chemistries can lead to erroneous conclusions about the epoxy-surface bonding or interphase structure.

2.4.2 Physisorbed Material

The surface chemical groups present on the adherend surfaces consist of species chemically bonded to the surface. A real surface exposed to the environment contains not only the surface chemical groups but also physisorbed water, carbon monoxide and carbon dioxide molecules. The amount adsorbed depends on the surface free energy of the adherend and can range in thickness from portions of a monolayer to multilayers of material.

Drzal et al [34,43] have measured the amount and composition of material desorbed from the surface of carbon fibers up to temperatures seen by the fiber surface during composite processing. The amount and composition depends on the surface treatment used. (Fig. 5). Untreated fibers either of low modulus or intermediate modulus produced less than a monolayer of desorbed material at temperatures up to 150 °C but after surface treatment these same fibers increased the amount desorbed up to five fold. The composition of the desorbing material was mostly water with carbon monoxide an also. Although the amount of material seemed insignificantly small, the composition revealed that the constituents were volatile at the temperatures used in processing of the composites. The presence of these materials could be responsible for significant void content. Figure 7 is an example of the potential of these materials to disrupt the interphase. A single surface treated graphite fiber was not degassed before encapsulation in an epoxy matrix. The desorbed material formed very large voids around the fiber surface even though the amount of the material was less than three equivalent monolayers.

Since the glass surface is equilibrated with ambient air, water is present either in the physisorbed or chemisorbed state. This hydrated surface of glass is responsible for its reactivity. The cations present in the glass easily hydrate to provide a surface rich in hydroxyl groups. The ability of a hydroxylated surface to hydrogen bond with additional water molecules causes the formation of multilayers of water hydrogen bonded to the surface. Thermal treatment of the surface before bonding can remove some or all of the water and some of the hydroxyl groups depending on the desorbing temperature. The surface to be used in the reinforcement then would be expected to be basic in character due to the interaction of the alkali elements with water.

The polymeric reinforcing fibers while having surfaces which are less hydrophilic, can absorb water in bulk to about a percent by weight. This amount of material on a

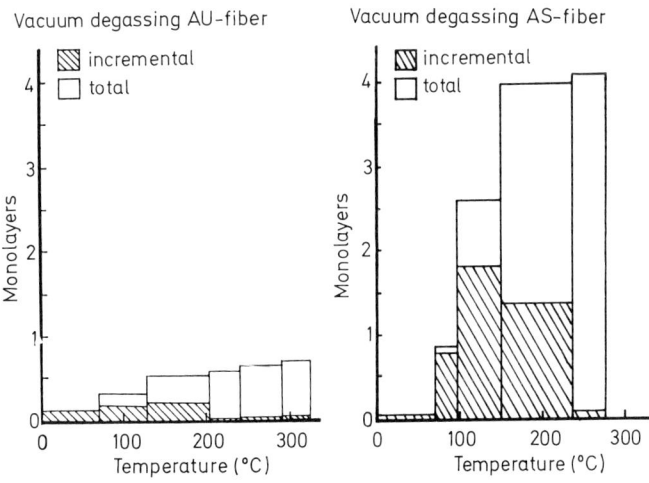

Fig. 5. The total and incremental amount, in equivalent monolayers, of volatile material desorbed from an untreated AU carbon fiber and a surface treated AS carbon fiber as a function of increasing temperature. From Drzal [34]

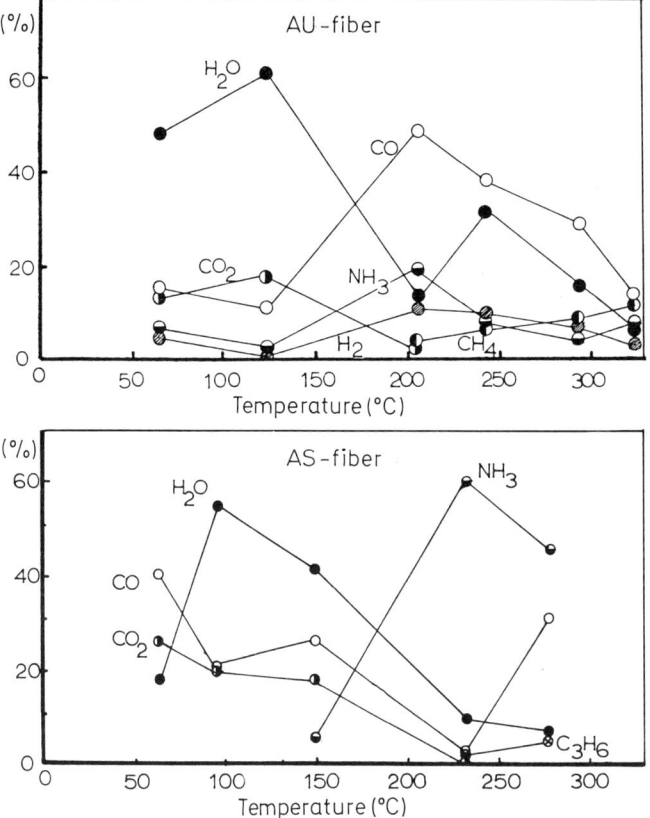

Fig. 6. The composition of material desorbing from an untreated AU carbon fiber and an AS surface treated carbon fiber as a function of temperature. From Drzal [34]

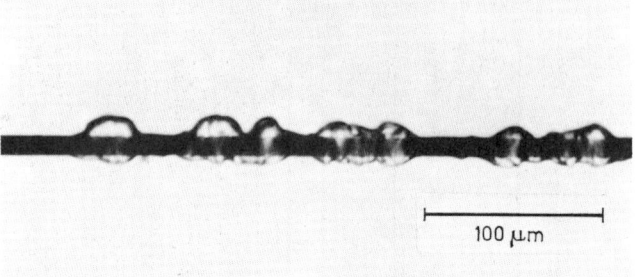

Fig. 7. An optical transmitted light micrograph of a single 8 micron carbon fiber in an epoxy matrix. The fiber had not been degassed before encapsulation. The volatile material on the surface vaporized around the time of gelation creating large voids around the fiber

volume basis is very small but concentrated at the interphase would provide a very large source of void forming material if it is not removed before processing. In all cases, care must be taken to remove these physisorbed species, either by direct gas phase desorption before processing or by solubilization in the epoxy during fabrication. The quantities involved are quite small on an absolute basis but locally they may exceed the maximum solubility in the epoxy and could have a pronounced effect on the local epoxy structure.

2.4.3 Topography

Topographical features comprise a significant portion of the interphase. In general, the epoxy matrix will conform to the topographical features of the substrate down to the molecular dimensions of the resin molecule. Since most epoxies are applied as a liquid of moderate to low viscosity, intimate contact between epoxy and substrate is achieved. Two aspects of the topographical features of the substrate must be considered as to their effect on the interphase structure of the epoxy.

The first of these is surface area. In fiber reinforced composites where the fiber diameters are on the order of 5 to 10 microns, at a 50 volume percent loading, the actual interfacial surface area can be on the order of 5,000 to 10,000 square meters per cubic centimeter of composite (Fig. 8). In adhesive bonds, a similar situation exists. A planar square centimeter of adherend surface can be covered with an oxide 100 to 200 nm thick having pores in close packed array of from 10 to 30 nm in diameter (Fig. 9). This produces a many-fold increase in interfacial area of contact between epoxy and adherend. Given a multicomponent reacting system, where the adsorptive potential for some of the reacting species is greater than others and where the free diffusion of species is retarded by coincident reaction, the potential for segregation of components at the interphase is very great when the actual interfacial area is large.

The mechanical effect of surface topography also exerts an influence on the interphase. The anodized oxides are the best examples of this feature. The oxides may be formed by a porous network 60 nm thick composed of pores close packed of 20 nm in diameter. The epoxy can penetrate into this structure to the bottom of the pores (Fig. 9). The effect is of a mechanical interlocking of the epoxy with the oxide which

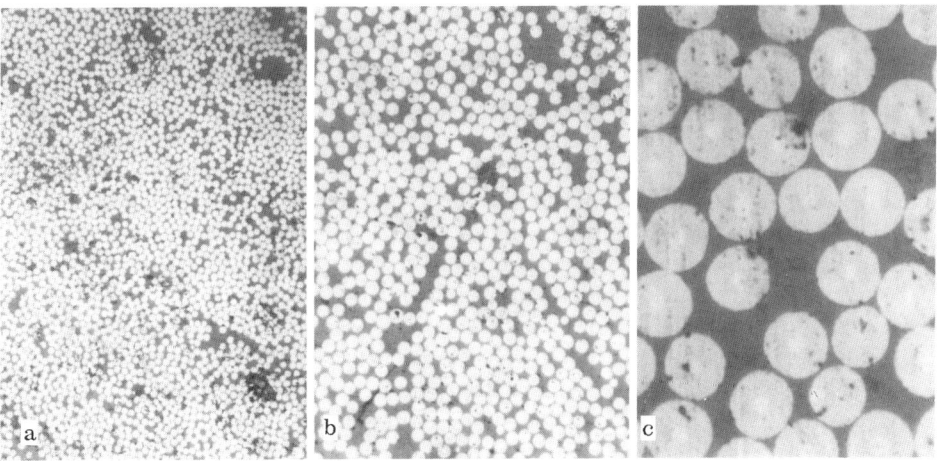

Fig. 8a–c. A polished refleacted light micrograph of a typical carbon fiber-epoxy composite showing the small amount of epoxy between fibers. Volume fraction 52%

Fig. 9. A transmitted electron micrograph of an ultramicrotomed section of an aluminum-epoxy interphase. The highly ordered structure in the center is a 3.3 micron thick aluminum oxide layer present on the base metal. The featureless area is the epoxy matrix. The light areas within the oxide are fractures caused by the microtoming. The epoxy has however penetrated to the bottom of all of the 50 nm pores in the oxide

resists both interfacial shear and tensile forces. Interfacial separation, per se, has no meaning since polymer and oxide can fracture under loading leaving behind a surface that is itself a composite of polymer and oxide.

2.4.4 Adsorption of Epoxy

The chemical and topographical nature of reinforcement and adherend surfaces presents an attractive surface for the adsorption of components commonly found in epoxy systems. The adsorption of polymers on surfaces has been known since the early work of Rebinder [44]. Malinskii [45] published a review of polymer adsorption in general and some recent papers have been added to the literature which document the adsorption of epoxy components themselves. Sergeyeva et al. [46, 47] have measured the adsorption of different molecular weight epoxy resin molecules onto glass fibers and have concluded that the epoxy adsorbs to thicknesses of about 100 nm. Similar adsorption studies by Zakharychev et al. [48] on titanium oxide surfaces and on other metal oxides [49] indicate that adsorption layers of from 200–500 nm in thickness depending on the solvent and polymer concentration.

In a mixture of epoxy resin and curing agent, competitive adsorption may take place. The epoxy molecule may adsorb on the basal plane of carbon fibers [50]. The adsorbed epoxy may be displaced by a curing agent molecule resulting in a region where the local composition at the interface is different from the bulk [51]. Koutsky [52] has shown that in an amine cured epoxy system, for example, this type of segregation can occur. Amine preferentially adsorbs on a copper substrate after the mixed epoxy is brought into contact with the copper. Concurrently the reaction of the epoxy and the amine is taking place. A high concentration of amine occurs at the copper surface due to diffusion from the bulk resin in the immediate area near the surface. However, the redistribution of amine in the bulk can not readily take place because of the increase in viscosity due to the polymerization. The result is that a compositionally stratified epoxy is formed. A region of high amine concentration exists near the copper surface. Further away a from the surface, a region of lower amine concentration has formed due to the migration of the amine to the surface. An intermediate (i.e. stoichiometric) amine concentration exists in the bulk some hundreds of nanometers away from the actual copper-epoxy interface.

Compositional and morphological differences can exist in the interphase regions between reinforcement and matrix. The ability to predict 'a priori' the local composition or the local structure is not possible because of the complexity of this region and the multiplicity of components and reactions occurring simultaneously.

2.5 Interfacial Additives

2.5.1 Coupling Agents

In addition to preferential adsorption of epoxy components at fiber or adherend surfaces, chemically different species can be added to the interphase to improve or alter an interfacial property. Among the species, that can be added to the composite, coupling agents have a great effect on the interphase structure and properties.

Coupling agents are usually defined as chemical species that have two different functionalities at opposite ends of the molecule which can chemically couple with both the adherend surface and the epoxy. Most coupling agents in use today are silanes [53] although titanates have been recently introduced [54]. Three of the four substituents on the silicon molecule are reactive with the fiber surface through their silanol functionalities. The silanols react with surface hydroxyls to form oxane bonds. The remaining functionality on the silane is usually some group reactive with the polymer. For epoxy matrices, aminofunctional silanes are usually very effective [55]. The silanes as a class are very extensive and Plueddemann [56] discusses many specific applications for use with a variety of polymers and substrates.

The coupling between the surface of the reinforcement and the matrix and its effect on mechanical properties has been demonstrated by Ahagon and Gent [57] among others. But until recently the structure of the coupling agent has been thought of as a monomolecular layer at the epoxy-substrate interface [58]. Recent work has shown that the coupling agent itself forms a multilayer composed of different polymeric structure [59]. Koenig and Ishida [60] have probed the structure of coupling agents with infra-red spectroscopy and have detected differences in structure with position. They proposed that the first layer near the surface is chemically coupled to the fiber and laterally to other silane molecules forming a two-dimensional network. A second layer of some finite thickness consists of a three-dimensional crosslinked network of polymerized coupling agents. The third portion is an intermingled epoxy and coupling agent three-dimensional network. Boerio [61] has shown that this structure changes with the pH environment from which the coupling agent is applied as well as with the chemical structure of the adherend surface itself.

Because of the chemical and structural differences of this coupling agent interphase layer, the mechanical properties of this region would be expected to be quite different from the bulk epoxy. Indeed, Lipatov [62] has shown that the addition of silanes changes the mechanical strength and chemical resistance of interphase regions.

2.5.2 Finishes and Primers

Another chemical species which can be added to the interphase region in composites or in adhesive joints is the 'finish' or 'primer'. The 'finish' is usually applied to the fiber before composite fabrication or as a 'primer' to the adherend before joint formation. Both 'finish' and 'primer' are quite similar in that they consist of a resin rich layer of material applied to the surface to be brought into contact with the epoxy. This resin rich layer may be constrained to remain in the vicinity of the interphase and consequently alter the local structure of the epoxy.

Alteration of this epoxy structure is the result of the fact that the epoxy molecules are both reacting and diffusing at the same time. This process forms a concentration gradient with a high epoxy monomer concentration at the surface which gradually reduces to the bulk concentration away from the surface. The properties of an epoxy with an excess of resin can be quite different from the stoichiometric amount. Figure 2, for example, illustrated the alteration of cured epoxy mechanical properties with epoxy/amine ratio. Excess epoxy or less than the stoichiometric amount of amine produces a brittle material if the mixture is cured in the same manner as the stoichiometric amount (Fig. 2). The stoichiometric sample has the lowest modulus while excess amine produces increased brittleness. The potential for variation in local properties within the epoxy due to the presence of a 200 nm or less layer must be considered.

2.6 Wettability

The end result of the surface chemistry of the reinforcement, the adsorbed material, topographical features, and epoxy composition is in the formation of the polymerized epoxy on the reinforcement surface. In order for this to happen, the fluid epoxy mixture must be brought into contact with the reinforcement surface, wetting must take place and energy added to aid the polymerization. The wetting of the reinforcement by the epoxy is a necessary criterion for optimum mechanical properties.

Wetting can be described thermodynamically as the creation of an interface whose free energy decreases. If contact angles are used as a measure of wettability this means that the contact angle should be less than ninety degrees for wetting to occur. This happens when the surface free energy of the wetting liquid is less than that of the surface on which it is placed. For epoxies as a class of materials, the surface free energies are around 40 ergs per square centimeter. The amines have a surface free energy higher than the epoxy. Stoichiometric mixtures of epoxy and amine however, have surface free energies which are about the same value as the epoxy. This implies that the epoxy predominates on the surface or at the interface in an epoxy mixture [63].

The surface free energy of the solid reinforcement surfaces can not be measured directly but can be estimated through the measurement of contact angles with a variety of liquids of known surface free energy. Kaelble [64] and Drzal [29] have measured these values for carbon fibers used in advanced composites. In general, the surface free energies are such as to support thermodynamic wetting of the reinforcement surfaces by the epoxies. Results for glass and aramid fibers [65] are similar, although the difference between the aramid and epoxy surface free energy is small. The oxides generally are high in surface free energy and the porosity and microtopography present on their surface aid the wetting of the substrate.

3 Interphase Effects on Composite Strength

The great advantage of composite materials lies in the large strength to weight ratio that these materials achieve over conventional homogeneous materials. These high strength and stiffness properties occur mostly in the direction parallel to the fiber

FIBER + MATRIX = COMPOSITE

$$E_c = V_f E_f + V_m E_m$$

Fig. 10. Rule of Mixture method of predicting composite properties based on the component properties. From Tsai et al. [66]

axis. Therefore, composite structure is usually made from plies or laminae of unidirectional material stacked one on top the other and varied in fiber axis orientation with respect to each other so that the resulting composite has an overall balance of properties in all directions. Even though this scheme places most of the load on the fibers themselves, some of the forces can nevertheless be applied in directions other than coincident with the fiber axis whereby the fiber-matrix interphase can be placed under a shear or a normal tensile load. Therefore, the interphase itself can influence composite properties.

The commercial composite materials being marketed today are optimized in order to make the interfacial properties acceptable in the sense that they will not fail at such low levels as to detract from the overall composite behavior. Considering a unidirectional specimen, where the fibers are all aligned parallel to each other, commercial systems can be described by a rule of mixtures [66] relationship (Fig. 10). Properties of the matrix and fiber can be linearly combined based on the volume fraction of each constituent. For example, the longitudinal tensile modulus is the sum of the proportion of each component. The interface in these systems is considered 'ideal' in that it efficiently transmits forces between fiber and matrix without failure. Using this model as a basis for micromechanical analysis and discussion, the magnitude of the forces present at the interface can be predicted.

3.1 Tensile and Compressive Strength

As a starting point for this discussion consider a composite specimen composed of unidirectional continuous reinforced lamina made of graphite fibers subject to a uniaxial tension in a direction parallel to the fibers [67] (Fig. 11). This composite lamina contains about 50% reinforcing fibers arranging in a hexagonally close packed network. Because of symmetry, we can isolate one fiber and its surrounding matrix and predict the forces acting at the fiber-matrix interphase, their intensity and direction based on the properties of the fiber and matrix. As the diagram of the enlarged fiber in its matrix shows, periodicity around the fiber combined with symmetry make analysis of a 30 degree segment only required for complete analysis of the cylindrically symmetric fiber.

Application of a tensile stress takes place in the fiber direction. The resultant stresses can then be calculated referenced to this applied longitudinal stress (σ_l). The radial stresses (σ_r) vary from slightly compressive to slightly tensile around the fiber. The magnitude varies with the volume fraction (V_f) of fibers. A high volume fraction of fibers means very little matrix between fibers. The shear stresses (τ) are quite low in magnitude but greater than the radial stresses. The circumferential or hoop stresses (σ_c) are largest. Although the magnitude of these stresses seems low compared to the

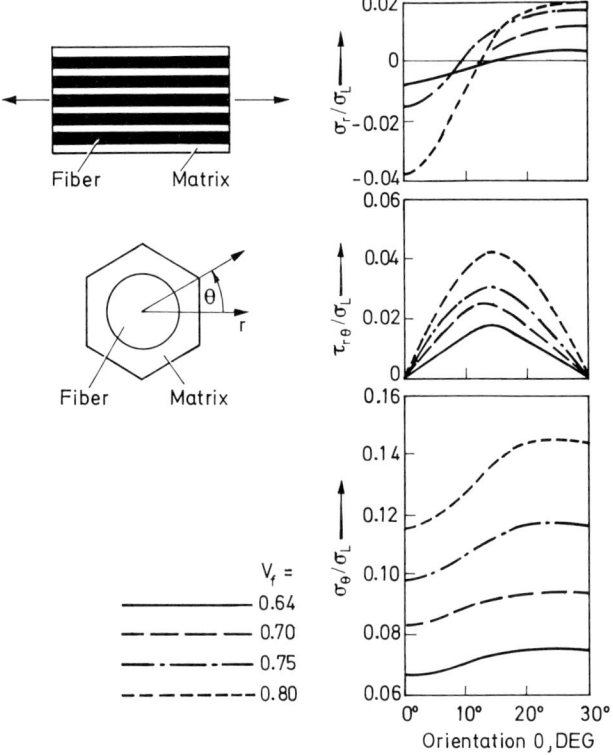

Fig. 11. A unidirectional lamina under longitudinal tension. The micromechanical analysis of the radial, shear and hoop stresses show an increase with fiber volume fraction. From Haener et al.[67]

applied stress, there is a large difference between fiber and matrix properties which can make even these apparently small stresses significant. For example, with an applied longitudinal load of 1400 MPa which is less than the tensile strength of most graphite fibers, the analysis would predict 124 MPa as the circumferential stress at the interface. The yield stress of a typical structural epoxy is about 100 MPa. Therefore even under longitudinal loading the limits of the epoxy matrix could be exceeded.

If the direction of the applied load is rotated by 180 degrees, the forces on the composite become compressive. Composite compressive strength is very closely related to the tensile strength in magnitude except that the direction of the resultant forces is changed. Hahn[68] has recently investigated the compressive strain behavior of graphite fibers in matrices where both the stiffness and the adhesion between the fiber and matrix have been varied. The fiber properties dominate the composite behavior as in the tensile case. Curtis and Morton[69] have noted only a ten percent change in compressive strength with interphase alteration. The interphase itself does not appreciably affect the composite mechanical properties in the fiber direction.

Composite mechanical properties measured parallel to the fibers axis direction are not in general very sensitive to the interphase. The fiber itself is the major load bearing element in the composite and does not require much support from the matrix.

3.2 Transverse Strength

Real situations demand more uniformity in composite properties than can be provided by unidirectional composites. Therefore lamina stacking sequences are made where the fiber orientation is altered to provide good properties in all directions. Lamina composed of fiber and matrix in which the fibers are all parallel to each other are stacked on top of each other with a systematic variation in fiber direction. These lamina are then bonded together and the resulting material has more uniformity in properties. Likewise in short fiber or discontinuous fiber composites fiber orientation is random. Therefore properties in directions other than parallel to the fiber (i.e. off-axis) are important [70].

Consider the same unidirectional lamina with the stresses now applied perpendicular to the fiber axis as shown in Fig. 12. The local stress at the fiber matrix interface can be calculated and compared to the nominally applied stress on the whole lamina to give K, the stress concentration factor. The plot of the results of this analysis shows that the interfacial stresses at the point of maximum principal stress can range up to 2.6 times the applied stress depending on the moduli of the constituents and the volume fraction of the reinforcement. For a typical graphite-epoxy composite, with a modulus ratio of 70 and a volume fraction of 70% the stress concentration factor at the interface is about 2.4. That is, the local stresses at the interface are a factor of 2.4 times greater than the applied stress.

The assumptions used in the model to predict the stress concentration factor assumed homogeneity of the fiber and the matrix up to the fiber-matrix interface. This assumption is not correct in all cases. While glass may be homogeneous in all directions, graphite and aramid are not. The tranverse modulus of graphite varies with the tensile

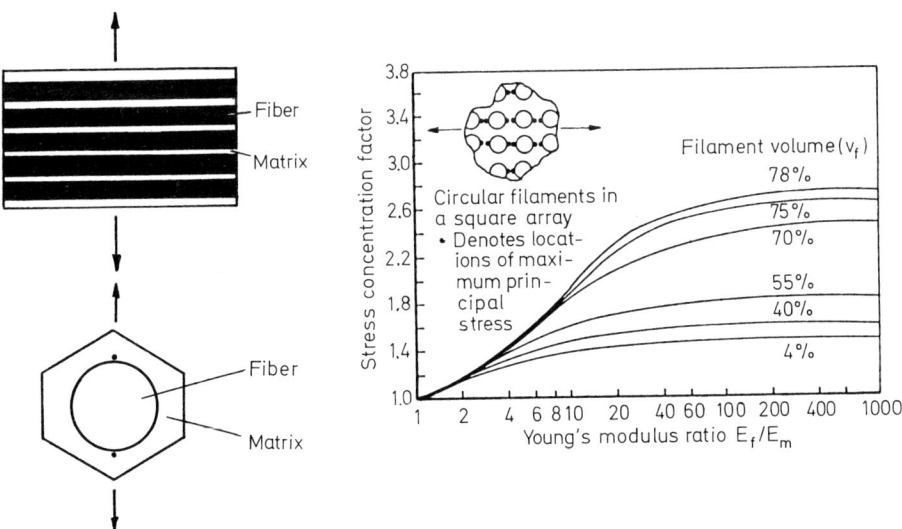

Fig. 12. A unidirectional lamina under transverse tension. The points of stress concentration are at the dots. The micromechanical analysis shows that the stress concentration factor increases with volume fraction of fiber and fiber to matrix modulus ratio. From Adams et al. [70]

modulus of the fiber which in turn is a function of the degree of alignment of the graphitic ribbons. Values for the transverse modulus can range from less than 5% of the longitudinal tensile modulus for the intermediate fibers to 15% for the lower modulus fibers [71]. The fibrillar nature of the aramid fiber makes the disparity between longitudinal and transverse properties of these fibers even greater (~4%).

On the epoxy side of the interface, high fracture toughness and low residual stresses [72, 73] are a requirement for optimum transverse strength in graphite and glass-epoxy [74] composites. Since the adsorption of epoxy components has been shown to be probable, the local structure of the epoxy at the interphase will most likely not be the same as in the bulk. This local anisotropy caused by the interphase is a limitation in the predictive capability of micromechanical models which do not include the interphase as a component.

3.3 Interfacial Shear Strength

Imposing a shear stress parallel to the fiber axis of a unidirectional composite creates an interfacial shear stress. Because of the disparity in material properties between fiber and matrix, a stress concentration factor can develop at the fiber-matrix interface. Under longitudinal shear stress as shown by the diagram in Fig. 13, the stress concentration factor is interfacial. The analysis shows that the stress concentration factor can be increased with the constituent shear modulus ratio and volume fraction of fibers in the composite. Under shear loading conditions at the interface, the stress concentration factor can range up to 11. This is a value that is much greater than any of the other loadings have produced at the fiber-matrix interface.

Interfacial shear strength is a critical property for composites. This previous analysis shows that the interfacial stresses under shear loading can be very large. Therefore as in the case for transverse strength, the interphase itself can be the controlling factor in the level of interfacial shear strength attainable for a given epoxy composite.

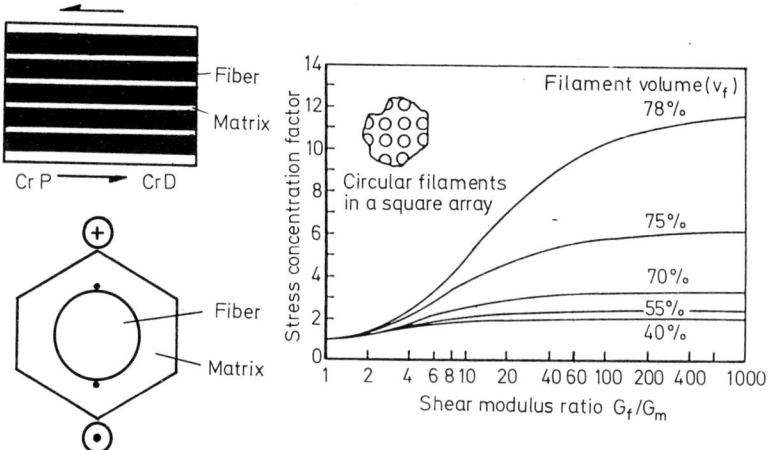

Fig. 13. A unidirectional lamina under shear loading. The dots indicate the points of high stress concentration at the interface. The micromechanical analysis shows that the stress concentration factor increases with volume fraction of fiber and fiber to matrix shear modulus ratio. From Adams et al. [70]

Although many thorough studies have been conducted with the goal of deriving structure-property relationships for epoxy systems, little success has been achieved in extending these relationships to the in-situ epoxy in the actual composite in such cases as interfacial shear strength for example. This can be explained by the lack of recognition of the dimensional changes operating in an actual composite and, therefore, the increased significance that a small interphase region can have when the fiber to fiber distance is only two to three times the thickness of the interphase itself. Drzal et al. [75] have investigated the mechanism of graphite fiber surface treatment and its relation to interfacial shear strength in an epoxy matrix.

Their results illustrate the necessity for including the interphase as a component of the composite and, therefore, as a factor in any materials behavior model of the composite.

Two different polyacrylonitrile precursor carbon fibers, an A fiber of low tensile modulus and an HM fiber of intermediate tensile modulus were characterized both as to their surface chemical and morphological composition as well as to their behavior in an epoxy matrix under interfacial shear loading conditions. The fiber surfaces were in two conditions. Untreated fibers were used as they were obtained from the reactors and surface treated fibers had a surface oxidative treatment applied to them. Quantitative differences in surface chemistry as well as interfacial shear strength were measured.

The results are plotted in Fig. 14. The upper two lines refer to the A fiber and the lower two lines to the HM fiber. For both fibers, the addition and removal of surface chemical groups did not produce reversible interfacial behavior. The untreated fiber surfaces produced results that could not be duplicated when the surface groups were removed. Microtoming of single fiber specimens pinpointed changes in the locus of interfacial fracture that were relatable to the interphase conditions caused by the surface treatment.

The main effect of the surface treatment is the removal of the native defect containing surface on the fibers. This surface layer itself can not sustain any great degree of

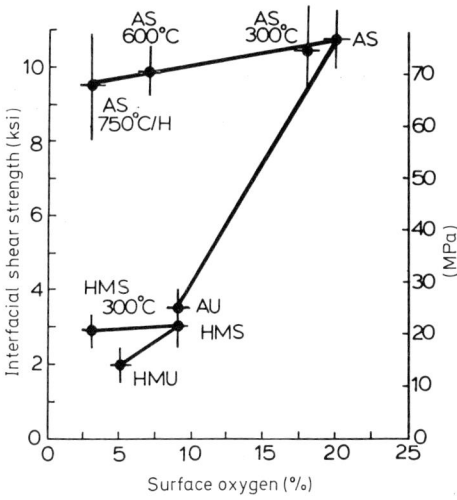

Fig. 14. The interfacial shear strength versus surface oxygen concentration for the A and HM carbon fibers. The large increase from the untreated (U) case to the surface treated case (S) for both fibers is due to removal of the native defect surface layer plus the addition of surface chemical groups. Removal of the surface groups with various treatments indicates that the surface chemical effect was a minor part of the overall increase. From Drzal et al. [75]

shear loading without failure. Once it is removed however, the defect free surface remaining can function mechanically at higher shear loadings. The addition of surface chemical groups by themselves add additional interaction of the epoxy with the fiber but only about 10% of the total improvement caused by the surface treatment. The HM fiber results parallel those of the A fiber except that the magnitude of the effect of the surface treatment on interfacial shear strength is reduced. This is due to the fiber structure. The much higher degree of fibril orientation on the outer part of the fiber makes basal plane shear failure more probable and therefore provides an upper limit to the interfacial shear strength.

The major conclusion from this work is that the interphase structure itself was the major factor in controlling interfacial shear strength and that the surface treatments altered the interphase structure.

Another example of interphase effects on interfacial shear strength is modification of the polymer side of the interphase through the application of a 'finish'. Kardos [76, 77] first showed that the inclusion of a small layer of material of different properties could have dramatic consequences when he placed elastomeric innerlayers between glass fibers and an epoxy matrix. He found that the composite short beam shear strength increased. Drzal et al. [78] placed 100–200 nm layers of pure epoxy around graphite fibers and measured the interfacial shear strength for both the surface treated fibers without a 'finish' and the surface treated fibers with this 'finish'. The interfacial shear strength for four different types of carbon fibers increased when an epoxy rich 'finish' layer was used (Fig. 15). A major observation was that the mode of interfacial failure changed when the 'finish' layer was present. Investigation of the interphase layer with model compounds showed that an amine deficient layer formed around the fiber resulting in an epoxy that had a high modulus but which was very brittle. The higher modulus increased stress transfer and was responsible for an increase in interfacial shear strength but the resulting interphase layer was low in fracture toughness.

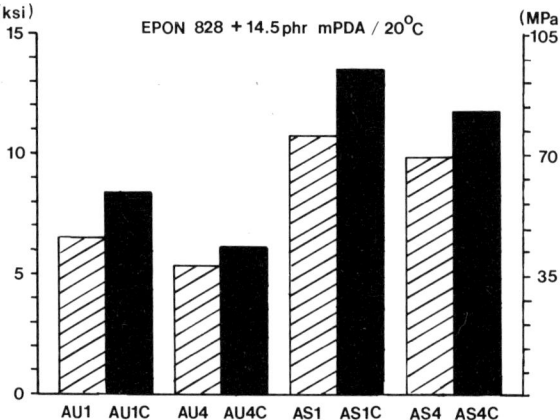

Fig. 15. The interfacial shear strength for coated fibers (C) versus uncoated carbon fibers in an epoxy matrix. The finish layer increases the shear strength by creation of a brittle interphase. From Drzal et al. [78]

Matrix cracks would grow at fiber breaks. When this amine deficient layer was not present, the failure was interfacial.

This alteration of the interphase by the addition of an epoxy rich layer caused a change in the fundamental material behavior of the interfacial zone and consequently caused significant changes in the composite behavior as well.

4 Interphase Effect on Composite Fracture

Structural applications of composite materials require not only acceptable static mechanical properites but the ability to withstand the generation and propagation of cracks without premature failre. For example, impact resistance, fracture toughness and fatigue resistance are desireable composite properties. Fiber-matrix structure at the interphase can affect the values attainable for these properties.

The effect of the interphase on composite fracture can best be shown by hypothetically separating the fracture process into its component parts to determine which are interphase dependent [79-83].

A fracture propagating perpendicular to the fibers can be considered as having many energy dissipative component mechanisms all of which add up to provide the fracture 'toughness' of the material.

4.1 Fiber-Matrix Debonding

For example, consider the crack tip as it intersects a fiber (Fig. 16). The local stresses at the tip can cause fiber-matrix debonding. The crack tip continues to open causing the interfacial debonded region to extend. The fiber continues to interact with the matrix through a frictional sliding force even after the initial bond fails. The distance over which the force acts is the debonded length times the difference in strain between the fiber and the matrix.

As Fig. 16 shows, the work on a per fiber basis can be calculated as a function of the interfacial shear strength (τ), the fiber diameter (d), the debonded length (l_d) and the difference in strain ($\Delta\varepsilon$).

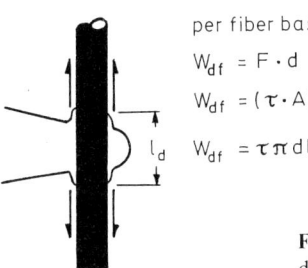

per fiber basis

$W_{df} = F \cdot d$

$W_{df} = (\tau \cdot A)(\Delta \varepsilon l_d)$

$W_{df} = \tau \pi d l^2 \dfrac{\Delta \varepsilon}{2}$

Fig. 16. Interaction of a crack with a fiber through fiber-matrix debonding

4.2 Fiber Deformation and Fracture

In addition to the fiber debonding step, energy is stored through deformation and fracture of the fiber (Fig. 17). This work can be calculated for each fiber as the product of the fiber tensile strength (σ), diameter and debonded length divided by the fiber tensile modulus (E_f). If an interfacial shear strength exists between fiber and matrix this relationship is altered by equating the debonded length to the critical transfer length (k_c).

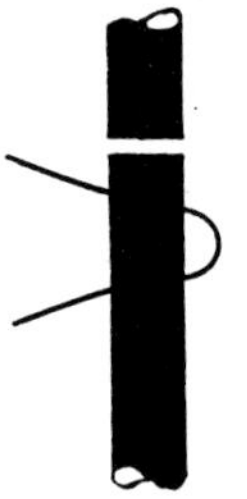

per fiber basis

$$W_{ff} = (1/2) \sigma_f^2 A \varepsilon l$$

$$W_{ff} = (1/8) \frac{\sigma_f^2 \pi}{E_f} d^2 l_d$$

$$W_{ff} = \frac{\sigma_f^2 \pi d^2 l_d}{8 E_f}$$

Fig. 17. Interaction of a crack with a fiber through fiber deformation and fracture

4.3 Fiber Pull-out

A third mechanism operating at the crack tip is the energy required to pull the broken fiber out of the matrix (Fig. 18). The work required is the product of the shear strength times the pull-out length (l_p). The total work of fracture can now be estimated as the sum of each of the contributions of the above mechanisms resulting in the relationship

$$W_t = \tau \pi d l_d^2 \frac{\Delta \varepsilon}{2} + \frac{\sigma_f^3 \pi d^3 l_d}{32 \tau E_f} + \tau \pi d l_p^2$$

in which it now becomes obvious that, the interfacial shear strength is a component of each of the terms that contribute to the total work of fracture. It is important to observe that the interfacial shear strength is not a linear function and, therefore, there will be an optimum condition of interfacial shear strength for maximum work. Although this model ignores the dissipative work done by the matrix, it does give some insight into the effect of fiber-matrix adhesion and, therefore, interphase structure on the work of fracture.

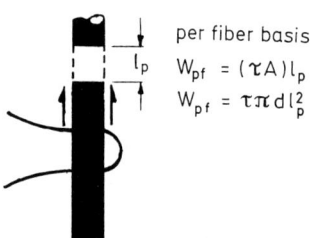

per fiber basis

$$W_{pf} = (\tau A) l_p$$

$$W_{pf} = \tau \pi d l_p^2$$

Fig. 18. Interaction of a crack with a fiber through fiber pull-out

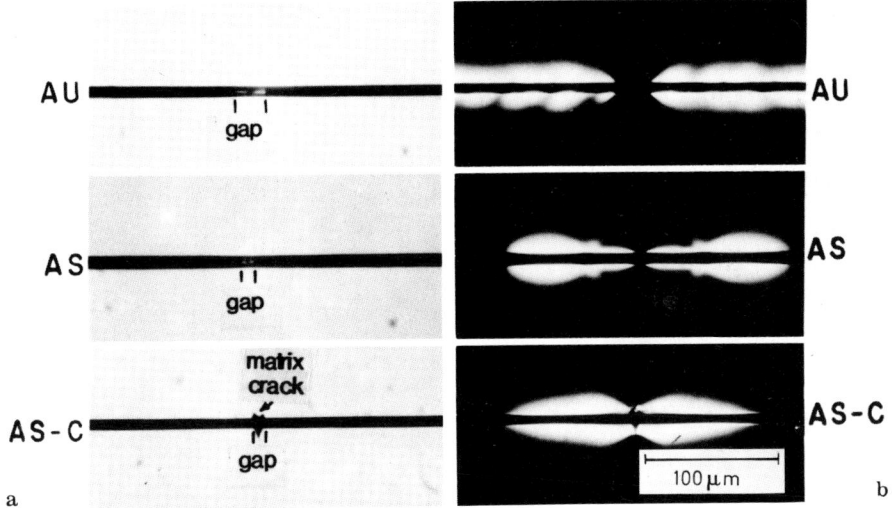

Fig. 19. Transmitted and polarized light micrographs of a single graphite fiber in an epoxy matrix under axial loading. The fiber and matrix are the same. Only the interphase has been changed. Three different failure modes develop as a result of the interphase alteration. From Drzal et al.[84]

Some experimental results support the general trends predicted here. Drzal et al.[84] have examined the differences in fracture toughness perpendicular and parallel to the fiber direction in graphite fiber-epoxy composites. The material variables were kept constant but the interphase was altered to provide three levels of interfacial shear strength (Fig. 19). Low levels were obtained by using an untreated fiber which would debond from the matrix by failure in the outer layers of the fiber. Medium interfacial shear strength was obtained when that same fiber was surface treated. The interfacial shear strength increased by two and one-half times and the failure mode became interfacial. After the application of a 'finish' layer, the level of interfacial shear strength increased again by 25%. The failure mode also changed from interfacial to matrix. Mini-compact fracture specimens were fabricated and tested. The micrographs of the fracture surfaces of these combinations for fracture perpendicular to the fiber axis are shown in Fig. 20. The interphase with low adhesion gives a large degree of fiber pullout which is a low energy dissipative process. Intermediate levels of adhesion produce a surface which has a moderate degree of pull-out. High levels of adhesion produce a surface which is almost planar. Although quantitative measurements were not reported based on the previous model analysis, the intermediate condition would be expected to produce the highest degree of fracture toughness because of the combination of fiber and matrix fracture, fiber debonding and fiber pull-out with a high degree of interfacial shear strength.

Fracture was also conducted parallel to the fiber direction in the same study with the same materials. Micrographs of the fracture surfaces of these composites are shown in Fig. 21. For low levels of adhesion, many bare fibers are detected. As the level of adhesion increases after the fibers are surface treated, both bare and epoxy covered surfaces are detected. An increase in the fracture toughness is measured with

Fig. 20. Scanning Electron Micrographs of the fracture surfaces of epoxy composites made with the same A carbon fiber with three different interphase conditions. Fracture is perpendicular to the fiber axis. From Drzal et al. [84]

this condition. For the finished fibers where the interfacial shear strength is greatest, all fracture paths appear to follow a path through the epoxy matrix and in between the fibers. Under this type of loading the brittle interphase does not cause the poor fracture behavior but instead forces the crack path to meander between and around fibers leading to increased fracture toughness.

Qualitatively fracture toughness translates into resistance to fracture propagation. In a composite as with any real material, fibers have a distribution of strengths rather than one failure strength for each fiber. This means that some fibers will break at low values of stress and can cause some local damage. Fracture paths can then proceed in either of two directions depending on the degree of fiber-matrix adhesion. If the interface is weak, the fibers will debond from the matrix and the fracture path will branch upward along each fiber-matrix interface. If the interface is very strong, the fracture path will propagate across the fibers and the sample. Each condition is undesireable. Debonding of the fibers would reduce composite behavior to that of a bundle of independent fibers. No shear transfer would take place and the fiber reinforcing effect would not be seen. Perpendicular crack growth is also undesireable since crack growth across the fibers would be catastrophic and premature failure of the composite would result. Behavior in between these two extremes would confine the damage to a region around the broken fiber without noticeable loss in composite properties [85].

The conclusion to be made from this work is that the fiber and its properties as well as the epoxy matrix and its properties are the same for all three cases. Only the interphase has been altered. Strength of materials fracture models would not predict a difference in fracture toughness and yet experimentally alteration of a 200 nm interphase zone changes the composite fracture properties dramatically.

FRACTURE SURFACES OF EPON 828 / mPDA / A Fiber COMPOSITES

Fiber	AU1	AS1	AS1C
Vol.Fract.	0.59	0.52	0.55
τ (MPa)	28	84	102

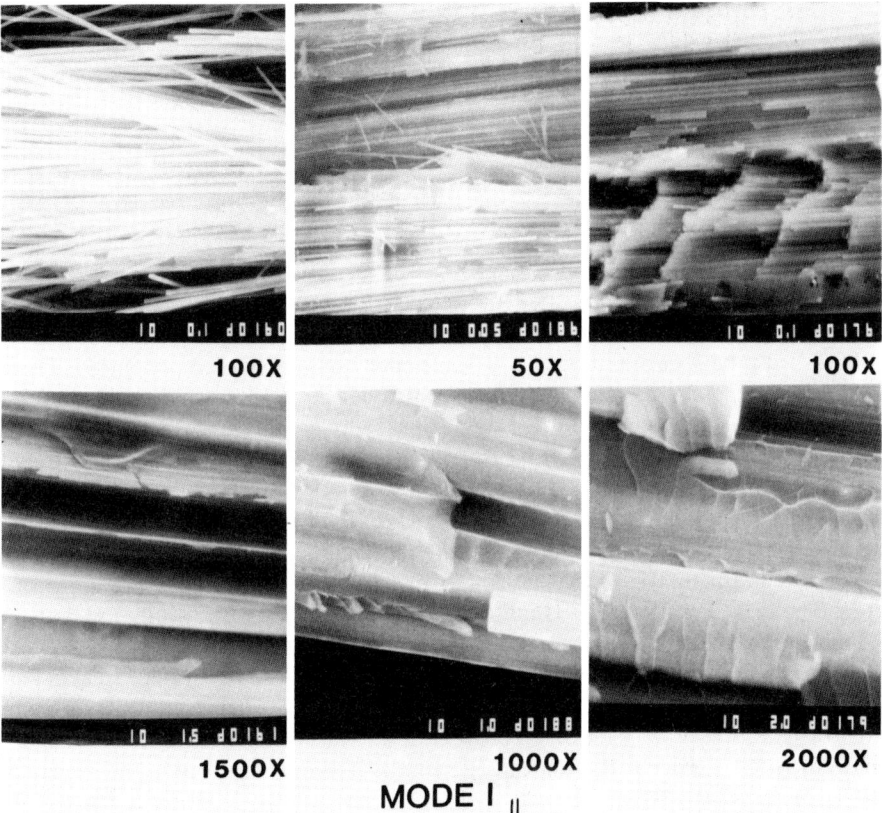

MODE I / II

Fig. 21. Scanning Micrographs of the fracture surface of epoxy composites made with the same A carbon fiber with different interphase conditions. Fracture is parallel to the fiber axis. From Drzal et al. [84]

5 Interphase Effects on Composite Environmental Resistance

The environmental resistance of epoxy composites has come to mean the ability to withstand elevated temperature moisture exposure. Many studies have been completed on these composites [86–89] and the major conclusions were that the epoxy matrix absorbs the major portion of the moisture with the result being a reduction in the epoxy matrix T_g and, therefore, a reduction in the upper operating limit of the composite. If the absorption and desorption of moisture is done at equilibrium conditions, the plasticization of the matrix is reversible. However, there is usually a significant

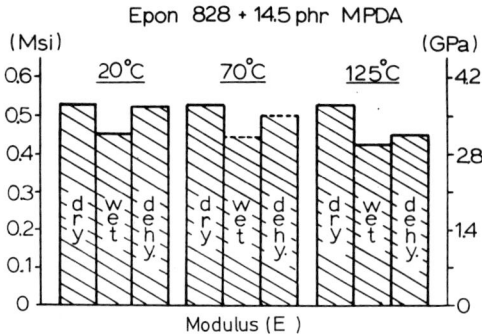

Fig. 22. The reduction in the initial tensile modulus of an epoxy after saturation with moisture and after dehydration under reversible conditions at 20, 70 and 125 °C

irreversible portion associated with this process. These irreversible changes have been proposed to be associated with changes at the interphase.

Drzal et al.[90] have investigated the effect of interphase modification on interfacial moisture absorption. The fibers used were a surface treated and a surface treated and finished type A carbon fiber in the same epoxy matrix studied previously. Three equilibrium exposure conditions were investigated. 20 °C, 70 °C and 120 °C were selected for moisture equilibration of single fiber samples and for the neat epoxy resin. The interfacial shear strength was measured both in the saturated and the dehydrated cases and compared to the initial dry values.

Figure 22 is a plot of the initial tensile modulus of the epoxy matrix after equilibrium moisture exposure and dehydration. At both 20 °C and 70 °C, the effect of moisture absorption on the matrix is reversible as evidenced by the reattainment of dry properties. The exposure at 125 °C is not completely reversible as shown by the data.

The effect of moisture exposure on the interfacial shear strength is shown in Fig. 23 for exposure at 20 °C. For both the surface treated fiber and the surface treated and finished fiber, the exposure to moisture causes a reduction in interfacial shear strength. After dehydration of the sample, a recovery of interfacial shear strength is noted but not to the full level of the dry unexposed sample. If stress is applied in the wet state, a permanent irreversible loss in interfacial shear strength is noted. Although the ab-

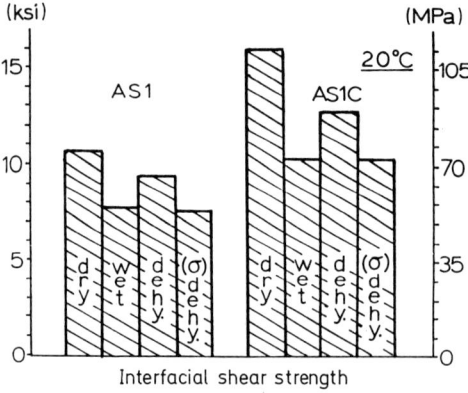

Fig. 23. The interfacial shear strength of a surface treated (ASl) and a surface treated and coated (ASlC) fiber after saturation with moisture at 20 °C (WET), dehydration at 20 °C (DEHY) and after being stressed wet and then dehydrated at 20 °C (DEHY)

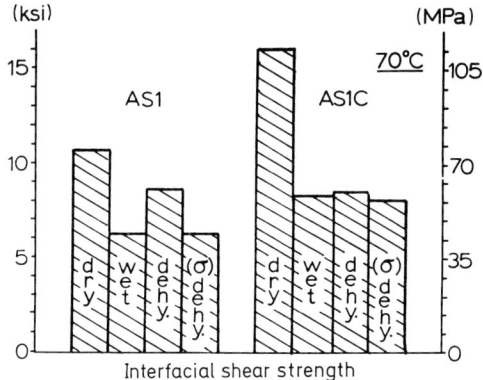

Fig. 24. The interfacial shear strength of a surface treated (ASI) and a surface treated and coated (ASIC) fiber after saturation with moisture at 70 °C (WET), dehydration at 70 °C (DEHY) and after being stressed wet and then dehydrated at 70 °C (DEHY)

solute magnitudes are different, the relative changes in interfacial shear strength are about the same for both the surface treated and the surface treated and finished fiber.

Exposure to 70 °C gives similar results for the surface treated fiber (Fig. 24). That is, a complete reversibility in noted. The finished fiber (i.e. the fiber with the interphase consisting of the amine deficient brittle interlayer) experiences a nonrecovery of interfacial shear strength after moisture exposure and dehydration. Parallel surface spectroscopic investigation of the fiber surfaces show that under these conditions the fiber surface chemistry is not permanently altered by this exposure. Model studies of epoxies with the amine deficient composition of the interphase show that the wet T_g of this material is about 70 °C. Therefore, the interphase is at or above its wet T_g and therefore because of the compliant nature of this material, stresses cannot be transfered efficiently and the interface is permanently distorted.

Exposure at 125 °C is very severe for this epoxy matrix (Fig. 25). Permanent changes in the matrix are noted. The interphase layer, however, acts to mitigate some of the deleterious interfacial effects and allows that system to regain a larger portion of its interfacial shear strength after moisture exposure and dehydration. The fiber without the finish layer has lost almost all of its interfacial shear strength and recovers very little after dehydration.

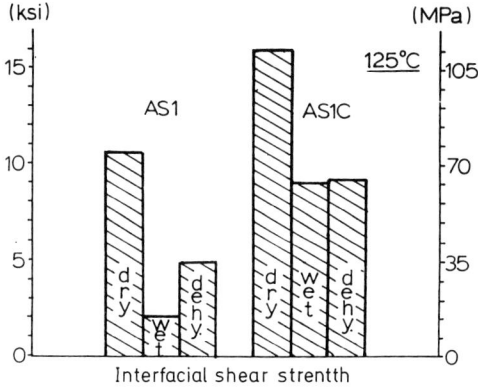

Fig. 25. The interfacial shear strength of a surface treated (ASI) and a surface treated and coated (ASIC) fiber after saturation with moisture at 125 °C (WET), dehydration at 125 °C (DEHY)

The major conclusion is that again although the nominal bulk properties of fiber and matrix are the same for the two systems investigated, the presence of a small interfacial layer of different composition had very large effects on the interfacial properties.

6 Conclusions

Composite components, both fiber (or adherend) and matrix, have chemical, morphological and structural variability that can be operative at or near surfaces to form an interphase. Bulk characterization of these materials ignores these components because of the dilution of their effect in the bulk. However, when composites are fabricated, fiber to fiber distances are on the order of tenths of microns. Interphase structures are also on the order of tenths of microns and can be a significant proportion of the structure of the material between fibers.

The interphase in epoxy composites is an important material component and can have significant effects on over all composite performance. It is not a fiber (adherend) or matrix property but it is a product of the interaction of fiber and matrix. Its existence has been the subject of speculation primarily because commercial materials are optimized systems which have minimized the deleterious effects of an interphase and analytical models of composite behavior based on empiricle material properties artificially ignore it.

Given the existence of interphases and the multiplicity of components and reactions that interact to form it, a predictive model for 'a priori' prediction of composition, size, structure or behavior is not possible at this time except for the simplest of systems. An in-situ probe that can interogate the interphase and provide spatial chemical and morphological information does not exist. Interfacial static mechanical properties, fracture properties and environmental resistance have been shown to be grealy affected by the interphase. Careful analytical interfacial investigations will be required to quantify the interphase structure. With the proper amount of information, progress may be made to advance the ability to design composite materials in which the interphase can be considered as a material variable so that the proper relationship between composite components will be modified to include the interphase as well as the fiber and matrix (Fig. 26).

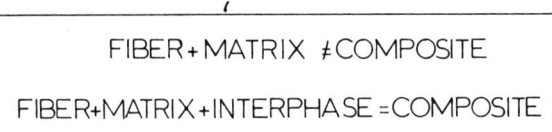

Fig. 26. Design Equation when the interphase is included as a material system variable

7 References

1. Agarawal, B. D., Broutman, L. J.: Analysis and Performance of Fiber Composites, New York Wiley-Interscience, (1980)
2. Sharpe, L. H.: J. Adhesion 4, 51 (1972)
3. Drzal, L. T., Rich, M. J., Lloyd, P. F.: J. Adhesion 16, 1 (1983)

4. Lee, H., Neville, K.: Handbook of Epoxy Resins, Chapter 1, New York McGraw-Hill, (1967)
5. Ibid., Chapter 4
6. Ibid., Chapter 5
7. Lee, H.: ASTM Bulletin, July (1960)
8. Hadad, D. K., Fritzen, J. S., May, C. A.: Exploratory Developments of Chemical Quality Assurance and Composition of Epoxy Formulations, Air Force Materials Laboratory Technical Report, AFML-TR-77-217, (1978)
9. Gutfreund, K., Kutscha, D.: Interfacial Investigations in Advanced Fiber Reinforced Plastics, Air Force Materials Laboratory Technical Report AFML-TR-67-275, 59 (1967)
10. Selby, K., Miller, L. E.: J. Mater. Sci. *10*, 12 (1975)
11. Kim, S. L. et al.: Polym. Eng. Sci. *18*, 1093 (1978)
12. Gupta, V. B. et al.: Polym. Eng. Sci., (in press)
13. Byrne, C. A., Hagnauer, G. L., Schneider, N. S.: Polym. Composites *4*, 206 (1983)
14. Morgan, R. J., O'Neal, J. E.: J. Mater. Sci. *12*, 1966 (1977)
15. Takahama, T., Geil, P.: Makromol. Chem., Rapid Comm., *3*, 389 (1982)
16. Mijovic, J.: J. Appl. Polym. Sci. *25*, 1179 (1980)
17. Mijovic, J., Koutsky, J. A.: Polymer *20*, 1095 (1979)
18. Aspbury, P. J., Wake, W. C.: Brit. Polym. J., *11*, 17 (1979)
19. Swetlin, B. J.: Fracture of Highly Crosslinked Epoxy Networks, Ph.D. Dissertation, University of Akron, (1984)
20. Bell, J. P.: J. Appl. Polym. Sci., *27*, 3503 (1982)
21. Fitzer, E., Geigl, K.-H., Huttner, W., Weiss, R.: Carbon *18*, 389 (1980)
22. Donnet, J. B., Bansal, R. C.: Carbon Fibers, New York Marcel Dekker, (1984)
23. Ehrburger, P., Donnet, J. B.: Carbon *15*, 150 (1977)
24. Diefendorf, R. J., Tokarsky, E.: Polym. Eng. Sci. *15*, 150 (1975)
25. Badanir, D. V., Joiner, J. C., Jones, G. A.: Nature *215*, 386 (1967)
26. Johnson, J. W., Rose, P. G., Scott, G.: Proc. 3rd Conf. on Ind. Carbon & Graph., Soc. Chem. Ind. London, 443 (1971)
27. Donnet, J. B.: Carbon *6*, 161 (1968)
28. Boehm, H. P., Diehl, E., Heck, W., Sappock, R.: Agnew. Chem. *3*, 669 (1964)
29. Hammer, G. E., Drzal, L. T.: Appl. Surf. Sci. *4*, 340 (1980)
30. Hopfgarten, F.: Fibre Sci. Tech. *11*, 67 (1978)
31. Brewis, D. M. et al.: Fibre Sci. Tech. *12*, 41 (1979)
32. Ishitani, A.: Carbon *19*, 269 (1981)
33. Proctor, A., Sherwood, P. M. A.: Carbon *21*, 53 (1983)
34. Drzal, L. T.: Carbon *15*, 129 (1977)
35. Rynd, J. P., Rostogi, A.: J. Coll. Interf. Sci. *54*, (1976)
36. Avakian, P., Blume, R. C., Gierke, T. D., Yang, H. H., Panor, M.: Polym. Preprints, Amer. Chem. Soc. *21*, 8 (1982)
37. Allred, R. E.: Surface Chemical Modification of Polyaramid Filaments with Amine Plasmas, Ph.D. Dissertation, Mass. Inst. Tech., Dept. of Mater. Sci., (1983)
38. Penn, L., Larsen, F.: J. Appl. Polym. Sci. *23*, 59 (1979)
39. Venables, J. D.: J. Mater. Sci. *19*, 2431 (1984)
40. Baun, W. L.: J. Mater. Sci. *15*, 2749 (1980)
41. Baun, W. L.: J. Adhesion *7*, 261 (1976)
42. Textor, M., Grauer, R.: Corrosion Sci. *23*, 41 (1983)
43. Drzal, L. T., Meschner, J. A., Hall, D.: Carbon *17*, 375 (1979)
44. Rebinder, P. A., Margaritov, V. B.: Zh. Rezin. Prom. *12*, 991 (1936)
45. Malinskii, Y. M.: Russ. Chem. Rev. *39*, 704 (1970)
46. Sergeyeva, L. M., Todosiychuk, T. T., Fabulyak, F. G.: in Naukova Dumka, p. 79 (1974)
47. Sergeyeva, L. M. et al.: Naukova Dumka, Geter. Poli. Mater., 96 (1973)
48. Zakharychev, V. P., Shchelkanova, T. S., Yermilov, P. I.: Naukova Dumka, Geter. Poli. Mater., 110 (1973)
49. Intorre, B. J., Kwei, T. K., Peterson, C. M.: J. Phys. Chem. *67*, 55 (1963)
50. Ko, Y. S., Forsman, W. C., Dziemanowicz, T. S.: Polym. Engr. Sci. *22*, 805 (1982)
51. Horie, K., Murai, H., Mita, I.: Fibre Sci. Tech. *9*, 253 (1976)

52. Racich, J. L., Koutsky, J. A.: Boundary Layers in Thermosets, Chemistry and Properties of Crosslinked Polymers, ed., S. Labana, 303, Academic Press (1977)
53. Yeung, P., Broutman, L. J.: Polym. Eng. Sci. *18*, 62 (1978)
54. Monte, S. J., Sugarman, G.: Proc. 35th Ann. Tech. Conf. SPI, Paper 23-F (1980)
55. Vaughan, D. J., McPherson, E. L.: Proc. 27th Ann. Techn. Conf. SPI, Paper 21-C, (1972)
56. Plueddemann, E. P.: Silane Coupling Agents, New York Plenum Press (1982)
57. Ahagon, A., Gent, A. N.: J. Polym. Sci. *13*, 1285 (1975)
58. Vogel, B. et al.: Proc. 22nd Ann. Tech. Conf. SPI, Paper 13B (1967)
59. Ishida, H., Koenig, J. L.: Polym. Engr. Sci. *18*, 128 (1978)
60. Ishida, H., Koenig, J. L.: J. Coll. Interface Sci. *64*, 565 (1977)
61. Boerio, F. J. et al.: Proc. 38th Ann. Tech. Conf. SPI, Paper 4-C (1983)
62. Lipatov, Y.: J. Adhesion *10*, 85 (1979)
63. Weaver, F.: Epoxy Adhesive Surface Energies Via The Pendant Drop Method, Air Force Materials Laboratory Report, AFWAL-TR-82-4179 (1982)
64. Kaelble, D. H., Dynes, P. J., Cirlin, E. H.: J. Adhesion *6*, 23 (1974)
65. Penn, L. S., Bystry, F. A., Marchionni, H. J.: Polym. Composites *4*, 26 (1983)
66. Tsai, S. W., Halpin, J. C.: Effects of Environmental Factors on Composite Materials, Air Force Materials Laboratory Report, AFML-TR-67-423, (1969)
67. Haener, J., Ashbaugh, N., Chia, C. Y., Teng, M. Y.: Investigation of the Micromechanical Behavior of Fiber Reinforced Plastics, USAAVLABS-TR-66-66
68. Hahn, H. T., Williams, J. G.: Compression Failure Mechanisms In Unidirectional Composites, NASA TM 85834 (1984)
69. Curtis, P. T., Morton, J.: Progress in Science and Engineering of Composites, ICCM-IV, Tokyo (1982)
70. Adams, D. F., Doner, D. R., Thomas, R. L.: Mechanical Behavior of Fiber Reinforced Composite Materials, Air Force Materials Laboratory Report, AFML-TR-67-96 (1967)
71. Tsai, S. W., Composites Design-1985, Think Composites, Davton, Ohio (1985)
72. Lee, S. M., Schile, R. D.: J. Mater. Sci. *17*, 2066 (1982)
73. Lee, S. M.: J. Mater. Sci. *19*, 2278 (1984)
74. Joneja, S. K.: SAMPE Quarterly, July, 31 (1984)
75. Drzal, L. T., Rich, M. J., Lloyd, P. F.: J. Adhesion *16*, 1 (1983)
76. Kardos, J. L.: J. Adhesion *5*, 119 (1973)
77. Kardos, J. L.: Trans. N. Y. Acad. Sci., Series II *35*, 136 (1973)
78. Drzal, L. T. et al.: J. Adhesion *16*, 133 (1983)
79. Wells, H. K., Beaumont, P. W. R.: J. Mater. Sci. *17*, 397 (1982)
80. Harris, B., Beaumont, P. W. R., deFerran, E.: J. Mater. Sci. *6*, 238 (1971)
81. Harris, B., Morley, J., Phillips, D. C.: J. Mater. Sci. *10*, 2050 (1975)
82. Outwater, J. C., Murphy, M. C.: Proc. 24th Ann. Tech. Conf. SPI, Paper 11-C, (1969)
83. Mullin, J. V.: Analysis of Test Methods for High Modulus Fibers and Composites, 349, ASTM STP 521, Philadelphia, PA (1973)
84. Drzal, L. T., Rich, M. J.: Research Adv. in Comp. in the US and Jap., ASTM STP 864, Philadelphia, Pa (1985)
85. Goan, J. C., Martin, T. W., Prescott, R.: Proc. 28th Ann. Tech. Conf. SPI, Paper 21-B (1973)
86. Kaelble, D. H. et al.: J. Adhesion *7*, 25 (1975)
87. Browning, C. E., Whitney, J. M.: Adv. Chem. Series No. 134, Amer. Chem. Soc., 137 (1974)
88. McKague, E. L., Halkias, J. E., Reynolds, J. D.: J. Comp. Mater. *9*, 2 (1975)
89. Augl, J. M., Berger, A. E.: 8th Nat. SAMPE Tech. Conf., *8* (1976)
90. Drzal, L. T., Rich, M. J., Koenig, M. F.: J. Adhesion *17*, 49 (1985)

Editor: K. Dušek
Received April 24, 1985

Epoxy Adhesion to Metals

Randall G. Schmidt and James P. Bell
Institute of Materials Science, U-136, University of Connecticut,
Storrs, CT 06268/USA

Metal/epoxy structural adhesive bonding systems and coating systems possess the potential to provide strong metal-to-metal bonds, with a number of distinct advantages over conventional metal joining techniques, and to successfully protect metals from damaging environments, respectively. Exceptional strength of adhesion can be achieved by these systems under dry conditions. However, because metal/ polymer adhesion systems generally exhibit a great reduction in strength in the presence of moisture, their use has been severely limited. This chapter reviews the major factors which influence the adhesion of epoxy resins to metals. Emphasis is placed on discussing the mechanisms by which water can decrease the strength of these systems along with the possible methods that can be employed to improve their durability in moist or wet environments.

1 Introduction 35

2 Metal/Polymer Adhesion System 36
   2.1 Metal Surface 36
   2.2 Metal Pretreatments 38
   2.3 Application of Epoxy Resins to Metal Substrates 40

3 Adhesion Strength-Dry Conditions 40
   3.1 Metal/Epoxy Adhesion Mechanisms 41
   3.2 Locus of Failure 42

4 Adhesion Strength-Wet Conditions 43
   4.1 Locus of Failure 43
   4.2 Strength Reduction Mechanisms 44
      4.2.1 Displacement of Epoxy by Water 45
      4.2.2 Oxide Layer Weakening by Hydration 46
      4.2.3 Corrosion-Induced Delamination of Epoxy-Based Coatings 47

5 Effect of Internal Stresses on Adhesion Strength 48

6 Methods Used to Increase the Durability of Metal/Epoxy Adhesion Systems 50
   6.1 Chemical Coupling Agents 50
   6.2 Formation of Metal Oxides Which Promote Mechanical Aspects of Adhesion 53
   6.3 Methods to Prevent Corrosion-Induced Delamination 56
      6.3.1 Decreased Water Permeation Through the Epoxy Coating 57

6.3.2 Decreased Oxygen Permeation Through the Epoxy Coating. . . . 58
6.3.3 Reduced Electrical Conductivity of the Oxide Layer 58
6.3.4 Incorporation of Cation-Exchange Materials into the Metal/Epoxy Interfacial Region 59
6.3.5 Use of Inhibitors 59
6.4 Relieving Internal Stresses 60
6.4.1 Addition of Fillers 60
6.4.2 Addition of Flexibilizers 61

7 Techniques Used to Determine the Locus of Failure 61
7.1 Ion Scattering Spectrometry (ISS) and Secondary Ion Mass Spectrometry (SIMS) 62
7.2 Auger Electron Spectrometry (AES) and X-ray Photoelectron Spectrometry (XPS) 64

8 Effect of Metal Identity 65

9 Conclusion 65

10 References . . . .

Abbreviations

AES	Auger Electron Spectrometry
D	Diffusion constant
EDX	Energy Dispersive X-ray Spectrometry
FPL	Forest Products Laboratories pretreatment
ISS	Ion Scattering Spectrometry
P	Permeability coefficient
PAA	Phosphoric Acid Anodizing pretreatment
SEM	Scanning Electron Microscopy
SIMS	Secondary Ion Mass Spectrometry
Tg	Glass transition temperature
WA	Thermodynamic work of adhesion required to separate two phases in an inert medium
WA_L	Thermodynamic work of adhesion required to separate two phases in water
WBL	Weak boundary layer
XPS	X-ray Photoelectron Spectrometry
σ	Actual fracture strength of adhesive bond
σ_i	Internal stress in an adhesive
σ_0	Fracture strength of stress-free adhesive bond

1 Introduction

The predominant applications of present day metal/polymer adhesion technology are for the development of strong metal-to-metal structural adhesive joints and durable protective coatings.

Adhesive bonding for structural joint formation is attractive because it presents a number of distinct advantages over more conventional metal joining techniques, such as [1-3]:
1) Bonding enables stresses to be distributed over large areas in the joint, thus avoiding the local stress concentrations present in riveted or spot-welded joints which can reduce fatigue resistance.
2) Bonding is often faster and cheaper than welding or bolting.
3) Bonding allows thin gauge metals and honeycomb assemblies to be fabricated, resulting in the availability of lighter structures.
4) Bonding permits the joining of dissimilar materials, and since adhesives are generally dielectric materials, their use minimizes the possibility of electrolytic corrosion when different metals are joined.
5) Bonding can eliminate crevices that often lead to crevice corrosion in riveted joints.
6) Bonding can greatly simplify design and construction techniques.

On the other hand, metal/polymer coating systems are of interest because polymers have the potential to protect metals from the expensive onslaught of corrosion (over 20 billion dollars annually are spent in the U.S. for materials to replace corroded items) [4]. Polymer coatings can protect metals by acting as barriers, thus preventing the formation of a complete corrosion cell and the spread of corrosion from an initial corrosion site [5].

Although many synthetic adhesives and surface coating formulations exist, this review will be concerned solely with evaluating the adhesion of epoxy resin based adhesives and coatings to metals. The authors feel that this restriction to epoxy resins is warranted because, in the areas of thermosetting adhesives and protective coatings, epoxies are generally considered as workhorse products [2,6]. There are a number of favorable characteristics that epoxy resins exhibit which have led to their popularity, such as [1,7]:
1) excellent adhesion to metals and many other substrates,
2) ability to be cured rapidly or slowly over a wide range of temperatures,
3) absence of water or volatile byproducts formed during cure reaction,
4) good wetting properties and low shrinkage during cure,
5) high level of mechanical strength,
6) outstanding toughness and chemical resistance
7) good electrical properties and thermal resistance.

The advantages and basic principles of structural bonding of metals with adhesives are widely known and accepted. However, only in the aerospace industry has this technology been used with success on a large scale. Adhesive bonding is important to this industry because it can be used to fabricate aluminum honeycomb sandwich structures with high strength-to-weight ratios for use in aircraft, space vehicles, rockets and missles [8]. The three major reasons why the full potential of structural adhesive joints has yet to be approached in other industries are, first, the durability of adhesive joints is poor in wet environments; secondly, a successful non-destructive testing method of the bonded joints does not exist; and thirdly, most of the structural adhesives currently used require elevated curing temperatures. The effectiveness of polymer coatings in protecting metals has also been severely limited by the substantial loss of adhesion strength in these systems under wet conditions. Therefore, in order for metal/polymer adhesive joint and protective coating systems to be of great value, their durability must be improved so that they exhibit good adhesion strength under all the environmental conditions to which they may be exposed.

This review will first present the metal/epoxy resin adhesive system and discuss how the presence of metal oxides influences adhesion; secondly, present the reasons why these systems exhibit very good adhesion strength under dry conditions and why this adhesion strength is greatly reduced in the presence of water; thirdly, examine possible methods of increasing their durability in wet environments; and finally, discuss some of the spectroscopic techniques that are currently being used to aid in the advancement of metal/polymer adhesion technology.

Since, in general, factors which influence the adhesion of metal/epoxy resin structural joints also influence adhesion in metal/epoxy resin protective coating systems, these two separate cases will often be combined and referred to simply as metal/epoxy adhesion systems.

Also, there exists an enormous number of different epoxy resin adhesive and coating formulations (many of which are proprietary in nature so that their composition has not been disclosed). The formulations can contain many different kinds of fillers and additives along with various types of epoxy prepolymers and curing agents. In fact, over fifty different chemical compounds can act as curing agents and convert the epoxy prepolymer into a crosslinked network. Similarly, there are many different types of epoxy prepolymers, all of which contain the characteristic three membered epoxy or oxirane ring in their structure [9]. Each formulation can result in an epoxy resin which exhibits different properties. In addition, by varying the ratio of the amount of epoxy prepolymer and curing agent used, the crosslink density and hence the rigidity of the resulting network can be varied [10]. Bell [11] has shown that although the tensile properties of epoxy resins are generally unaffected by varying the crosslink density at room temperature, the dynamic mechanical properties of the resin can be greatly affected. However, since this review is primarily concerned with discussing the major factors which influence the adhesion strength of metal/epoxy systems, the different epoxy resin formulations will not be discussed here (see Lee and Neville [9]). Therefore, in this review, the term epoxy resin will be used to collectively refer to all of the various epoxy resin formulations, and there may be some exceptions to the general principles that will be presented.

2 Metal/Polymer Adhesion System

2.1 Metal Surface

Based on the studies of adherend surfaces by Fowkes [12], Parks [13] and Zisman and Shafrin [14], Bolger [15] developed a model (Fig. 1) which depicts the fundamental characteristics of the hydrated oxide surface of any metal, metal oxide or silicate.

Epoxy Adhesion to Metals

Additional H₂O surface layers.
Thickness depends on temperature
and relative humidity.
Dots indicate hydrogen bonds.

First H₂O surface layers
tightly bound.
Surface hydroxyl groups.

Metal oxide layers.
Actual thickness and structure
depend on metal substrate.

Crystalline metal substrate.

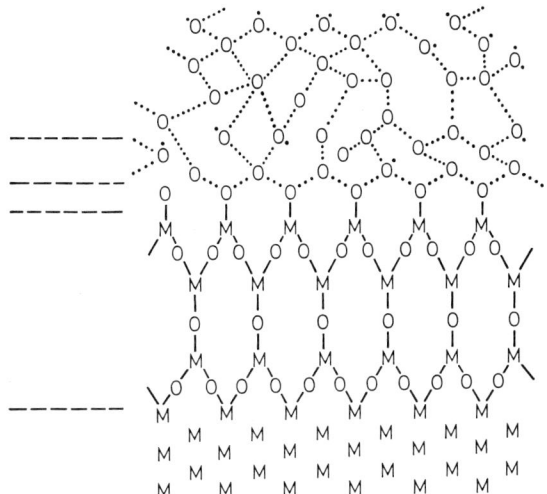

Fig. 1. Schematic representation of water and oxide layers on a metal surface. M = metal atom, 0 = oxygen, 0· = —OH, 0: = H₂O [15] (Reputed from Ref. 15, p.6, by courtesy of Plenum Press)

Upon exposure to oxygen, all metals form surface metal oxide layers which vary in thickness and structure, depending on the identity of the base metal and the oxide formation conditions. Mercury and noble metals generally form very thin oxide films. On the other hand, most metals of primary commercial importance (i.e. aluminum, iron, zinc, etc.), tend to form oxide layers which are thick enough (40–80 Å or more), so that the underlying metal atoms do not contribute in an appreciable way to the adhesion forces in metal/polymer systems [11].

The surface oxygens of the metal oxides hydrate to form surface hydroxyl groups under normal ambient conditions by way of the following reaction [15].

It has been reported that approximately one surface hydroxyl group per 50–100 Å² of surface area exists on most metal oxides [18,19].

From the above discussion, it might appear that an adhesive or polymer coating

---

1 For the case of copper, a mixture of cuprous and cupric oxides is present on the copper surface which acts as a defect semiconductor. Therefore, electrons can readily be transported from copper to its oxide surface allowing oxidation to continue at the metal oxide/adhesive interface [16]. This continued oxidation reaction which involves the base metal can interfere with adhesion between the oxide and the adhesive. Hence, the underlying metal atoms can effect the adhesion forces in some cases [17]

will adhere equally well to most metals since most have a hydrated oxide surface layer. However, this is not the case, because the activity of the hydroxyl groups is heavily influenced by the type of metal atom to which they are attached. Furthermore, the number of hydroxyl groups on the surface can be varied by changing the prebonding thermal history [15].

The presence of the hydrated oxide surface is advantageous for adhesion systems because it enhances the wetting of metal surfaces by epoxies and other polar resins. However, many interactions can take place between the environment and the hydrated metal oxide layer which can be detrimental to adhesive bonding. The presence of the hydrated oxide layer provides a surface on which water and polar organic contaminants can readily be adsorbed and retained. In fact, up to twenty molecular layers of water have been found to exist on 'dry' metal surfaces in studies of aluminum, copper and iron under normal ambient conditions [20]. On the other hand, non-polar contaminants are generally displaced from the metal oxide surface by polar contaminants, but they still can be present and interfere with bonding to some extent. Figure 2 shows the various layers of water and contaminants that can build up on metal surfaces and interfere with metal/polymer adhesion [21]. It is obvious that in order to prepare metals for optimum adhesive bonding they will have to be pretreated and dried to remove these weak layers [21].

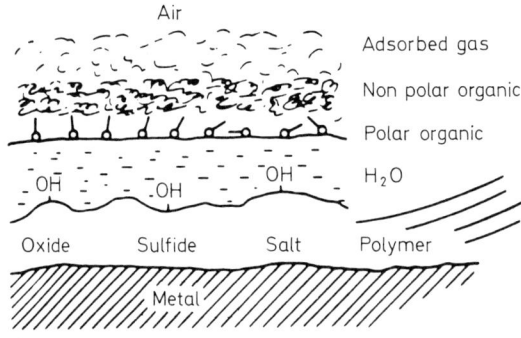

Fig. 2. Hierarchy of spontaneously adsorbed layers on a metal surface [21]. (Reprinted from Ref. 21, p. 56, by courtesy of Marcel Dekker, Inc.)

2.2 Metal Pretreatments

Prior to applying the epoxy resin or any other adhesive or coating, the metal substrate is always pretreated in an attempt to provide a surface to which the resin can strongly adhere. Metals that are available for industrial applications are normally covered with a thick contaminant layer (see Fig. 2), which has properties that differ greatly from those of the bulk metal. This layer may consist of mill scale, processing lubricants, water, or various other contaminants from the atmosphere which can adsorb on the high energy metal surfaces. Because the contaminant layer is usually very thick, it generally controls the forces that develop between the metal surface and the adhesive or coating. A solvent degreasing pretreatment is designed to remove some types of surface contamination. Therefore, this should be the minumum pretreatment employed prior to adhesive bonding [22].

There are numerous metal pretreatment techniques that can be used in addition to solvent degreasing. To obtain metal/polymer systems which exhibit strong *initial* adhesion it is usually sufficient to wash the metal with a solvent followed by an acid etch or sandblasting technique to remove any weak oxide layers and roughen the surface simultaneously.

Using an acid etch or a mechanical abrasion technique, it is possible to completely remove an oxide layer from a metal surface. However, since new oxide reforms almost instantaneously, it is impossible to have an oxide free surface present under practical bonding conditions. On the other hand, by controlling the environment in which the new oxide layer is generated, the thickness and structure of the oxide can be somewhat regulated [21]. This is important because, in order to produce metal/polymer adhesive systems which are durable, it is necessary to have stable oxides which are receptive to the adhesive [1]. Great advances have been made in forming aluminum oxides which exhibit increased stability in water and also have the ability to enhance adhesion by a mechanism involving mechanical aspects (the pretreatment and adhesion mechanisms will be discussed in section 6.2) [23].

Figure 3 [24] illustrates that the durability of metal/polymer adhesion systems can greatly be influenced by the metal pretreatment chosen [1]. Therefore, it is very important to select the best pretreatment for a given system.

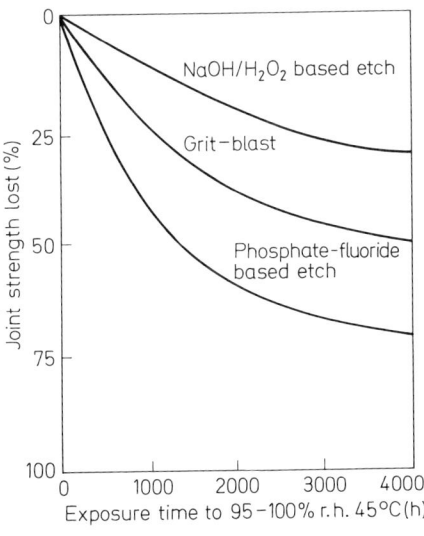

Fig. 3. Effect of substrate pretreatment on the durability of titanium/epoxy joints [24]

A summary of some of the more common metal pretreatments is given by Derjaguin [25]. The literature also describes many specialized pretreatments for steel [26-29], stainless steel [22,30], aluminum [23,31-35], copper [36-41] and other metals [42-44].

Allen and Alsalim [22] compared the effect of various pretreatments of stainless steel (martensitic structure) on the torsional shear strength of napkin ring joints formed with an epoxy adhesive (Redux 319 (Bonded Structures Ltd.)). They concluded that etching with any reducing acid will result in improved adhesive bond strength in this system beyond that which can be achieved with simple solvent degreasing methods. Furthermore, pretreating with either hydrofluoric acid or sulfuric acid was found to yield the best adhesion strength values. These acid etches remove the old oxide layers and new ones are produced which exhibit well defined macro- and micro-depressions that act as very good mechanical interlocking sites. They also reported that since a deposit of graphite remains on stainless steel surfaces following acid etching, post etch chemical desmutting with either nitric or chromic acid followed by drying should be completed to remove the weak graphite layer. The result is a surface which is favorable for adhesive bonding not only in terms of mechanical aspects, but also chemical aspects because hydroxyl groups, which can interact chemically with the polymer, will be exposed on the metal oxide surface.

Similarly, Allen, Alsalim and Wake [45,46] determined that alkaline hydrogen peroxide was the best pretreatment for titanium alloys. This pretreatment was found to preferentially etch the β phase, while also undercutting some of the α grains and redepositing needle-like crystals on the β grains. The very rough surfaces that resulted were found to enhance adhesion by mechanical aspects.

It has been shown that every step of a particular surface preparation may be of significance with respect to the resulting bond strength of an adhesion system [47]. Therefore, it is imperative that pretreatment procedures be followed exactly if the experimental results from various laboratories are to be rightfully compared.

2.3 Application of Epoxy Resins to Metal Substrates

Most epoxy adhesives used for structural bonding by industry are liquid or paste in form, with the epoxy and curing agent supplied as a two-component system (however, some single component 'latent cure' systems also are used). The two components are mixed (frequently elevated temperatures are used to enhance mixing) just prior to bonding, and then generally are applied to the metal substrate as a liquid. A solvent or mixture of solvents is often added to reduce the viscosity of the resin, particularly for coating applications, in order to ensure that sufficient wetting of the substrate occurs [9].

Epoxy based paints or surface coatings can be applied as liquids or solids to the metal substrates. Brushing, spraying or dipping are some of the most popular methods of applying liquid or solution epoxy resin/curing agent coatings. On the other hand, solid or powder single component epoxy resin systems are often applied to metal substrates via electrostatic or fluidized bed techniques [2].

A very wide range of curing times and temperatures exists for epoxy resin systems. Therefore, they offer a great deal of design flexibility for use both as structural bonding adhesives and protective surface coatings.

3 Adhesion Strength — Dry Conditions

High initial adhesion strength between epoxy resins and metals is readily obtainable as long as surface contamination and weak oxide layers have been removed from the metal surface by an appropriate pretreatment. Illustrative data will be presented below.

3.1 Metal/Epoxy Adhesion Mechanisms

The good initial adhesion strength values achieved with epoxy adhesives (aluminum alloy/structural epoxy single lap-joints, loaded in extension; typically 5–20 kN failing load) [48,49] and coatings (aluminum alloy/structural epoxy peel joints, 90° peel; typically 1–10 kN/m peel strength) [49] are primarily due to basic epoxy chemistry itself. Aliphatic hydroxyl and ether groups are present in the initial resin chain and in the cured polymer. Therefore, epoxy resins have high polarity. These polar groups serve as sites for the formation of strong electromagnetic bonding attractions (hydrogen bonds) between epoxy molecules and metal oxides (bond energy: 5–10 Kcal/mole) [50]. The importance of the epoxy hydroxyl groups in the adhesion of epoxy resins to aluminum is illustrated in Fig. 4 [51]. The epoxide group or oxirane ring can also aid in metal/epoxy adhesion through the formation of chemical bonds with active hydrogens on the metal surface [6].

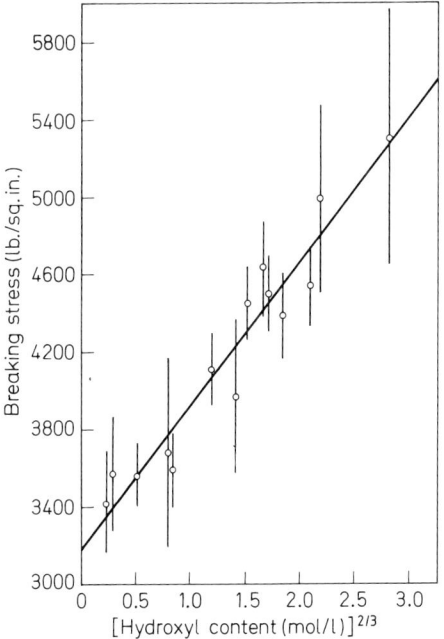

Fig. 4. The nominal breaking stress (Amsler testing machine) of aluminium alloy/epoxy single lap joints as a function of the hydroxyl content of the epoxy raised to the two-thirds power [51] (Reprinted from Ref. 51, p. 307, by courtesy of Society of Chemical Industry)

If the metal surface has been pretreated to provide a roughened surface, or a porous [52] or fibrous oxide [40,53] layer, then mechanical aspects can also play an important role in the adhesion strength of metal/epoxy systems. While the adhesive or coating mixture is liquid, prepolymeric epoxy resin and curing agent molecules can penetrate into the cavities and pores provided by the pretreatments. Upon curing the epoxy resin becomes mechanically embedded into the oxide structure. Bascom [54] and Packham [55] have suggested that when failure of these systems takes place, considerable plastic deformation of the embedded epoxy resin will occur, with pore and fibrous ends acting as nucleating sites for the deformation. The strength of an adhesive bond reflects the total amount of energy dissipated during failure [56]. Therefore, the energy dissipated in the plastic deformation of the resin will aid in promoting the adhesion strength of the system.

Dispersion forces [57], which result from temporary variations in the distribution of electron density in atoms, can account for up to 90 per cent [58] of the adhesion forces between non-polar polymers and metal substrates (bond energy: 0.5–5 Kcal/mole) [50]. However, for the adhesion of epoxy resins and other polar polymers to metals, dispersion forces are of secondary importance when compared to the electromagnetic and mechanical interactions discussed above.

3.2 Locus of Failure

Good durability of metal/epoxy resin systems can generally be achieved when these systems are limited to dry environment exposure [24,59,60]. Under dry conditions, no mechanism exists by which the strong interfacial forces between epoxy resins and metal substates can be destroyed. Also, since the forces across the interface are stronger than the cohesive properties of the epoxy resin itself (ultimate tensile strength 28–91 MN/m^2), failure under high stress invariably occurs cohesively within the resin [61–63]. Therefore, under dry conditions, the strength of a metal/epoxy system is usually governed by the cohesive strength of the epoxy resin. Due to the very good mechanical properties of epoxy resins, the strength achieved in these systems is more than sufficient for most applications.

The hot-dry desert site curve in Fig. 5 [61] illustrates the excellent durability that can be achieved with metal/epoxy systems under dry conditions. On the other hand,

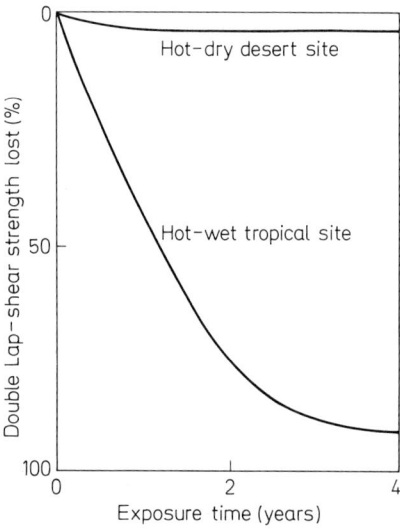

Fig. 5. Effect of outdoor weathering on the strength of aluminum alloy/epoxy-polyamide joints (chromic-sulfuric acid-etch metal surface pretreatment) [61] (Reprinted from Ref. 61, p.194, by courtesy of Gordon and Breach)

the hot-wet environment curve (Fig. 5) indicates that water is very detrimental to these systems. The reason for this great loss of strength in wet environments will be discussed in the following section.

4 Adhesion Strength — Wet Conditions

All adhesion scientists will agree that water is a very destructive environment for metal/polymer adhesion systems (see Fig. 5). Since water is one of the most common environments encountered, the effectiveness of metal/polymer coating and structural bonding systems has been severely limited by this great loss of adhesion strength in the presence of water.

4.1 Locus of Failure

Under dry conditions, failure of metal/polymer bonded systems usually occurs cohesively within the resin. However, upon prolonged exposure to water, failure is generally found to occur interfacially between the polymer and the substrate (i.e. adhesive failure) [60-65]. As illustrated in Fig. 5, this change in the locus of failure is accompanied by a large reduction in adhesion strength. Therefore, water is believed to reduce the strength of adhesion by reducing the strength of the interfacial region.

Kerr, MacDonald and Orman [66] completed a study of the change in cohesive strength of an epoxy adhesive and the change in shear strength of aluminum/epoxy joints upon exposure to water and to ethanol at 90 °C. The results of this experiment are shown in Fig. 6 [66]. They found that ethanol had a much larger effect, when compared with water vapor, in decreasing the cohesive strength of the epoxy adhesive

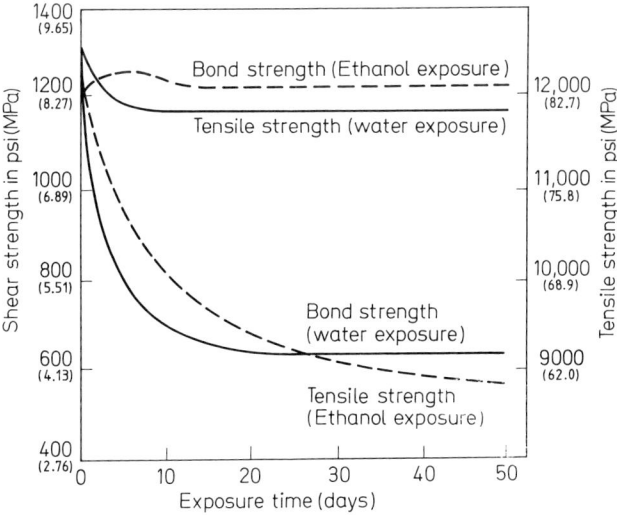

Fig. 6. Shear strength of aluminum/epoxy joints and tensile strength of epoxy resin after exposure to water and ethanol at 90 °C [66] (Reprinted from Ref. 66, p. 63, by courtesy of Society of Chemical Industry)

alone. On the other hand, the water vapor had a much larger detrimental effect on adhesive joint shear strength. These results add support to the belief that the decrease in joint strength upon exposure to water is attributable to a decrease in the strength of the interfacial region rather than the bulk adhesive.

In addition, it has been shown that the decrease in the cohesive strength of epoxy resins in water is due to plasticization and is completely reversible [67,68]. Kerr and coworkers [66,69,70] and others [71,72], however, have shown (Fig. 7) [66] that although the effect of water on joint strength is largely reversible, it is at least partially irreversible.

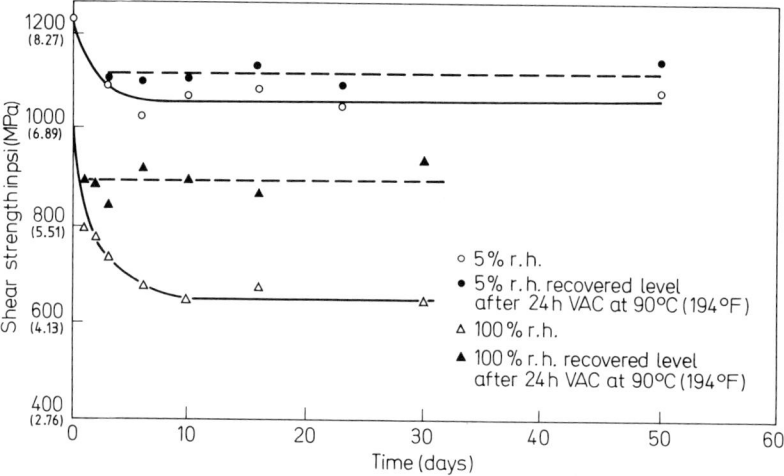

Fig. 7. Shear strength of aluminum/epoxy joints after exposure to humid and non-humid environments and after post-exposure drying at 90 °C [66] (Reprinted from Ref. 66, p. 64, by courtesy of Society of Chemical Industry)

4.2 Strength Reduction Mechanisms

Metals have very high surface energies (500–700 mJ/m²) [73,74] and all organic adhesives and coatings are permeable to water to some extent. The combination of these two properties results in a situation in which it is virtually impossible to prevent water from reaching the interfacial region of metal/epoxy adhesion systems under wet conditions. Water can enter either by diffusion through the bulk epoxy or it may be transported along the metal/epoxy interface. Brewis, Comyn and Tegg [64] observed an inverse linear relationship between aluminum/epoxy joint strength and the water content of the joints. They concluded from their study, and others agree [32,75,76], that water generally enters a joint by diffusion through the epoxy, rather than by passage along the interface. This is, of course, likely to depend upon the type of epoxy, the thickness, and the temperature.

Once water reaches the interfacial region in sufficient quantity (the critical concentration of water in an adhesive has been reported to be 1.35%) [77] it is generally agreed that the loss of strength of metal/epoxy systems is due to a reduction in strength of this interfacial region [60–65]. Numerous mechanisms have been proposed in an attempt to account for this phenomenon. Although no one mechanism has been found which explains every case, there are a few which have been shown separately or collectively to successfully explain most of the behavior observed. These strength reduction mechanisms are discussed below.

4.2.1 Displacement of Epoxy by Water

Figure 8 shows schematically the strong hydrogen bonds in the metal/epoxy interfacial region which are believed to be one of the primary reasons for the large adhesion strengths that are observed under dry conditions. As discussed above, water will invariably reach the interfacial region under wet conditions. Since water molecules are very strong hydrogen bonding agents, they can readily break the bonds between the metal and the epoxy resin and form new hydrogen bonds with the hydrated oxide surface of the metal. The result is the displacement of the epoxy resin from the metal and the formation of a weak water layer at the interface. The presence of the weak water layer can greatly reduce the strength of the metal/epoxy system [66,78,79].

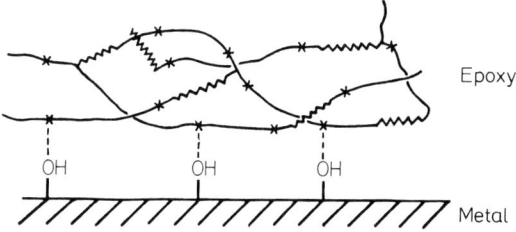

Fig. 8. Schematic representation of the hydrogen bonding that occurs between clean metal surfaces and epoxy resins (X = polar group, ——— = epoxy, ∿∿∿ = curing agent)

Kinloch, Gledhill and Dukes [75,80] investigated the interface stability by calculating the thermodynamic work of adhesion, W_a, required to separate the two phases in an inert medium and in water. Two examples of the calculated W_a (inert) and W_{aL} (water) values are shown in Table 1 for metal/epoxy systems [75,80]. The change in the work of adhesion values from positive to negative in the presence of water indicates that a driving force for the displacement of the epoxy exists. However, as Kinloch[1] noted, this method does not take into account interfacial forces arising from primary chemical bonds or mechanical aspects. Therefore, it cannot be used alone in predicting whether or not displacement will actually occur.

Table 1. Calculated values of W_a and W_{aL} for epoxy/metal oxide Interfaces [75,80] (Reprinted by courtesy of Gordon and Breach)

Interface	Work of adhesion	
	Inert medium, W_A (mJ/m^2)	In water, W_{AL} (mJ/m^2)
Epoxy/ferric-oxide	291	−255
Epoxy/aluminium-oxide	232	−137

Displacement of epoxy by water definitely plays an important role in the strength loss of metal/epoxy adhesion systems. However, there are also other mechanisms by which water can reduce the strength of the interfacial region.

4.2.2 Oxide Layer Weakening by Hydration

Water can reduce adhesion strength by reducing the strength of the metal oxide layer via hydration [52,81]. Hydration of the oxide layer is detrimental because the resulting aluminum-, iron-, or other metal-hydrates generally exhibit very poor adhesion to their base metals [52]. Therefore, the produced layer of hydrates will effectively act as a weak boundary layer in the system and decrease adhesion strength. Since the hydration reaction has been most heavily studied on aluminum oxides, the authors have chosen to base the discussion of the hydration mechanism on this case.

The structure of the oxide layers that are formed on aluminum can vary depending upon the pretreatment used [32,82]. However, before water is introduced, the bulk of the oxide layer in most cases exhibits amorphous Al_2O_3 morphology. Upon exposure to water, the initial step of the hydration is the conversion of Al_2O_3 to boehmite (A100H), as evidenced by x-ray photoelectron spectroscopy [83] and SEM [84]. The second step of the hydration consists of the nucleation and growth of bayerite ($Al(OH)_3$) crystallites. Ahearn, Davis, Sun and Venables [83], using high resolution SEM and x-ray diffraction analyses, have observed that the bayerite crystallites nucleate on the plates of the boehmite phase. However, it was not possible to determine whether a dissolution — redeposition or a nucleation mechanism was involved in the conversion of boehmite to bayerite.

Figure 9 [85] illustrates the adhesion strength loss mechanism involving the hydration of aluminum oxides which has been proposed by Venables and co-workers [85] based on crack propagation studies employing wedge testing. In humid environments the aluminum oxide is converted to a hydroxide which adheres very poorly to the substrate. The result is an increased propagation rate and a shift in crack position from the adhesive to the oxide/metal interface when changing from a dry to a wet environment.

Progress has recently been made at successfully retarding this adhesion strength loss mechanism for the case of aluminum oxide through the use of a phosphoric acid anodizing pretreatment or special inhibitors [86]. These pretreatments are analyzed in Section 6.2.

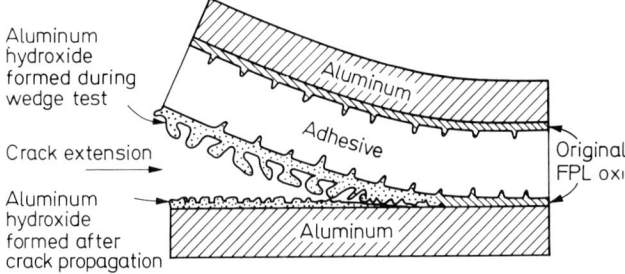

Fig. 9. Schematic diagram of the failure mechanism proposed by Venables et al. [85], in aluminum/polymer joint systems during wedge testing in humid environments. (Reprinted with permission from Chapman and Hall LTD.)

4.2.3 Corrosion-Induced Displacement of Epoxy-Based Coatings

Locus of failure studies [75, 80] on metal/epoxy joints that had been exposed to water indicate that corrosion of the metal substrate does not occur until after interfacial failure has occurred. This suggests that corrosion itself does not play a primary role in the loss of adhesion strength mechanism of metal/epoxy joints, but rather is a post-failure phenomenon. However, for the case of metal/epoxy protective coating systems, Leidheiser and coworkers [88, 91, 92] and Dickie and coworkers [5, 87, 89, 90] have proposed that localized corrosion processes are part of an important delamination mechanism.

In general, as long as good adhesion exists between the epoxy coating and the metal substrate, the metal will be protected from corrosion processes. However, if a defect or discontinuity exists in the epoxy coating which results in the metal being exposed to an electrolyte, an aqueous corrosion cell can develop. Delamination of the epoxy coating can be an indirect result of the electrochemical reactions that take place in the corrosion cell [91, 93].

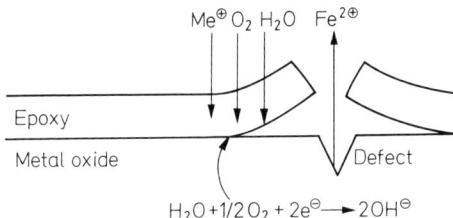

Fig. 10. Schematic diagram of the corrosion-induced delamination mechanism for a steel/epoxy coating system

Figure 10 shows schematically the primary corrosion reactions that are involved in the corrosion-induced delamination of an epoxy coating from a steel substrate [92]. The anodic reaction is the dissolution of iron ($Fe = Fe^{+2} + 2e^-$). This reaction can occur at a site where an aqueous medium is available to accept the ferrous ions which are produced. A defect in the epoxy coating provides such a site. The primary cathodic reaction is the formation of hydroxide ions by the reduction of oxygen ($H_2O + 1/2\,O_2 + 2e^- = 2\,OH^-$). There are five components that must be present for this reaction to occur, namely [91]: (1) water, (2) oxygen, (3) electrons, (4) positive counterions for the negative hydroxide ions generated, and (5) and oxide which is catalytically active for the reduction reaction. These components are generally present beneath the coating at the boundary of a defect in the presence of water. There are other cathodic reactions that can and do occur (i.e. hydrogen evolution, water formation, metal deposition) [92]. However, the production of hydroxide ions is of primary interest because these ions can cause alkali hydrolysis of the resin to occur. Indeed, it is this chemical degradation of the epoxy resin adjacent to the metal/coating interface that provides the pathway for delamination of the coating [5, 89, 94]. As the delamination proceeds the cathodic site moves, remaining just in front of the debonding region [92].

There are numerous methods with which one can attempt to reduce or prevent this corrosion-induced adhesion loss mechanism from occurring. Refer to Section 6.3 of this review for a discussion of these methods.

5 Effect of Internal Stresses on Adhesion Strength

Internal stresses are created in polymer coatings and adhesives during setting by the shrinkage of the polymer due to chemical changes (molecules moving from van der Waals distances to smaller covalent distances apart during cure reactions)[95] and physical changes (i.e. solvent evaporation)[96-98]. Also, internal stresses develop upon post cure cooling due to the differences in thermal coefficients of expansion between the adhesive and the substrate[99-101]. These stresses develop because the interfacial area of the coating or adhesive is forced to remain at its original size by adhesion forces between the polymer and the rigid metal substrate. Therefore, as the polymer approaches solidification and loses its flow properties, any further chemical, physical or temperature change will result in the formation of internal stresses in the system. The polymer will be able to relax a portion of the developed stresses. However, undoubtedly some internal stress will remain and "threaten the cohesive and adhesive properties of the system"[96].

Epoxy resin coatings cast onto metals from solution exhibit significant internal stresses. Shrinkage of the coating due to diffusion and evaporation of the solvent accounts for some of the stresses observed. In addition, since epoxy resins have coefficients of thermal expansion values up to ten times greater than that of a metal substrate[2], post cure cooling of the coating system to room temperature can add to the build-up of internal stresses[96,99].

Using an energy balance analysis, Croll[96] has developed a theory which has been shown to successfully predict the effect of internal stresses on the adhesion strength values of coating systems obtained from peel and pull-off tests. This theory requires the determination of the recoverable strain energy stored in the coating system from stress-strain data, and is limited to coating systems which exhibit adhesive failure. Employing this theory along with an additional theory which describes the dependence of internal stresses on coating thickness, Croll[102] has developed theoretical curves (Fig. 11) which illustrate the effect of epoxy coating thickness and solvent evaporation

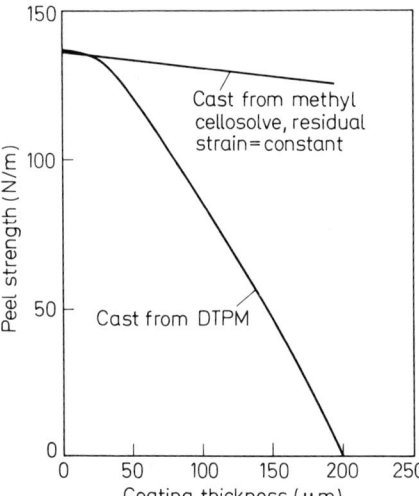

Fig. 11. Dependence of the peel strength of tin plate/epoxy systems on coating thickness for coatings cast from DTPM and methyl cellosolve. These theoretical curves were constructed using experimental values for modulus, critical coating thickness, solvent evaporation rate, solution concentration and interfacial work of adhesion[96] (Reprinted from Ref. 96, p. 123 by cautesy of Plenum Press)

rate on the peel strength of tin plate/epoxy systems. Internal stress increases and hence peel strength decreases with an increase in coating thickness. Also, Fig. 11 indicates that the use of a fast drying solvent (methyl cellosolve^R) will result in minimal internal stresses and hence superior peel strength values over those achieved using a slow drying solvent (tripropylene glycol monomethyl ether (DTPM)). A fast drying solvent is superior because most of the evaporation occurs while the coating molecules still have sufficient mobility, owing to the small extent of reaction, to relax the stresses formed due to solvent loss [102].

Solventless epoxy adhesives and coatings generally exhibit a much smaller internal stress build-up than their solution-cast counterparts. Shimbo, Ochi and Arai [99] formed various aluminium/solventless epoxy coating systems using a bisphenol-A-type epoxy resin cured with four different aliphatic α,ω-diamine curing agents at elevated temperatures. The internal stresses that developed in the systems were measured using a strain-gauge method. Their results indicate that for a solventless epoxy coating cured at elevated temperatures, the internal stresses developed are primarily due to the differences in the thermal expansion coefficients between the epoxy and the metal. Others [49] agree with this conclusion. Also, Shimbo et al. [79] have developed an equation based on the metal and epoxy thermal expansion coefficients and the tensile modulus and glass transition temperature (T_g) of the resin, which has been shown to provide a good estimation of the magnitude of the internal stress build-up. Figure 12 [99] shows experimental data which supports the theory of Shimbo et al. in that the degree of internal stresses present in solventless epoxy systems (cured at elevated temperatures) is dependent primarily upon the difference in temperature between the T_g of the epoxy and the use temperature, regardless of the epoxy network structure. Shimbo et al. have attributed these results to the very small amount of mobility that the network segments possess in the glassy region. Therefore, the excess shrinkage of the epoxy resin over that of the metal, due to its larger thermal expansion coefficient, is apparently converted directly to internal stress when the system is cooled below T_g. On the other hand, essentially no internal stresses are present at temperatures above T_g because the network segments have sufficient mobility to quickly relax them [99]. Dannenberg [103] determined that internal stresses of the order of magnitude 0.08 MPa K^{-1} are generated upon cooling alu-

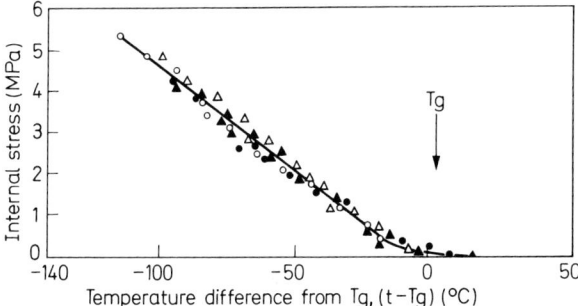

Fig. 12. Internal stress developed in epoxy resins upon cooling from above Tg to (t−Tg). (Different symbols represent data from the use of different curing agents) [99]. (Reprinted from Ref. 99, p. 49, by courtesy of the Federation of Societies for Coatings Technology)

minum/epoxy based coating systems below the glass transition temperature of the epoxy due to the excess thermal contraction of the epoxy.

In the past, the detrimental effect of internal stresses on adhesion strength has often been ignored. Conversely, in a few cases, the presence of internal stresses has been proposed as the primary reason for adhesive failure [96, 99, 104-108]. The authors believe that internal stresses can play an important role in reducing the adhesion strength of metal/epoxy systems and that the theories developed by Croll [96, 102] and Shimbo et al. [99] should lead to an increased awareness of this fact. In addition, further work must be completed in this area so that the usefulness of these theories and the effect of internal stresses on metal/polymer adhesion systems can be successfully analyzed.

6 Methods Used to Increase the Durability of Metal/Epoxy Adhesion Systems

In Section 4.2, the strength loss mechanisms of metal/polymer adhesion systems in the presence of water were discussed. From this discussion it is evident that high initial adhesion strength is not the only important property of these systems. Actually, if a metal/polymer adhesion system is exposed to humid environments, it is more important for the system to exhibit good durability.

Over the past fifteen years a number of different approaches have been taken in an attempt to increase the durability of the metal/polymer interfacial region in the presence of water. These attempts have met with varied degrees of success. However, to date adhesion scientists are still searching for a means of achieving sufficient wet environment durability, so that the enormous potential of metal/polymer adhesion systems can soon be utilized effectively. The authors have selected to discuss a few of the more promising durability-enhancing methods.

6.1 Chemical Coupling Agents

Coupling agents are generally low molecular weight multifunctional compounds which can chemically couple the polymer adhesive or coating to the metal substrate. They are normally applied to the metal substrate from solution as a final pretreatment step. Coupling agents possess the potential to form water stable covalent bonds across the metal/polymer interface which can greatly increase the durability of these systems. Gent [109], in a recent review of various chemical coupling agent studies [110, 111], concluded that chemical bonding at the interface can definitely act as a strengthening feature in adhesion systems. In addition, he reported that the most successful coupling agents have been those which are long extensible molecules [109, 112].

Silane coupling agents have been used since the 1950's to improve the bond between inorganic reinforcements and organic matrix resins in reinforced plastics [113]. More recently, Plueddemann [114] and others [29, 115, 16] have shown that silane coupling agents can also enhance the durability of metal/epoxy joint and coating systems. Table 2 shows the increased adhesion strength and durability that was achieved by

Table 2. Wet and recovered direct pull-off test adhesion strength values for aluminum/epoxy paint systems employing silane coupling agents [116]

Added % Silane/treatment[a]	Wet adhesion[b] MPa/% detached	Recovered adhesion[c] MPa/% detached
Degreased only	2.2/100	12.9/100
0.2% G/Degreased	26.3/0	27.3/20
0.2% F/Degreased	25.3/0	26.9/0
Sandblasted only	7.4/100	22.2/20
0.2% G/Sandblasted	25.1/0	26.7/5
0.2% F/Sandblasted	28.2/0	28.7/0

[a] Silanes: $Y-Si(OCH_3)_3$ F: Y = Diamine group,
  G: Y = Mercapto group;
[b] immersed in water 1500 hrs at room temperature;
[c] 48 hrs at room temperature and humidity

Walker [116] using silane coupling agents in aluminum/epoxy paint systems. Various silane coupling agents were also tested on steel, cadmium, copper and zinc. Using accelerated weathering exposure tests, it was found that several of the silanes could also enhance the adhesion strength and durability of these metal/epoxy paint systems [116]. The most common type of silane coupling agents used in metal/epoxy adhesion systems have the general structure $X_3Si(CH_2)_nY$, where n = 0 to 3, X is a hydrolyzable group on silicon and Y is an organofunctional group selected for optimum bonding to the epoxy adhesive or coating [62].

It is generally accepted that the increased durability observed is due to the successful formation of strong covalent bonds across the metal/epoxy interface. The hydrolyzable groups (X) on silicon can react with surface hydroxyl groups on the metal surface to form oxirane bonds (M—O—Si) [117, 118]. On the other hand, at the silane/epoxy interface the Y group reacts with hydroxyl and epoxide groups of the epoxy resin to complete the chemical couple. Silane coupling agent films have been found [119] to form a strong polysiloxane network which enhances their ability to increase the durability of the interfacial region.

Graham and Emerson [29] have developed a pretreatment for ferrous metals, utilizing silane coupling agents, which was shown to provide both high wet strength and corrosion protection in steel/epoxy adhesion systems. This pretreatment (SNS) involves first the deposition of a thin layer (about 100 Å) of tin hydrosol particles (corrosion inhibitor) on the steel substrate from a wetting hydrosol dispersion. This layer is then modified with an aqueous silane coupling agent solution. Figure 13 [29] illustrates the superior durability that has been achieved using the SNS pretreatment over that obtained using other popular pretreatments for steel/epoxy coating systems.

Although silanes are the most common coupling agents used for metal/polymer adhesion systems, other coupling agents [17, 65, 120-125] have been used with varied success. DeNicola and Bell have shown that both betadiketone [65, 123] and multifunctional mercaptoester [124] coupling agents can significantly increase the durability of steel/epoxy napkin ring joint [127] systems. Figure 14 [124] illustrates that very good durability was achieved in these systems for up to sixteen days immersion in hot water when a mercaptoester coupling agent was employed. The mercaptoester groups can

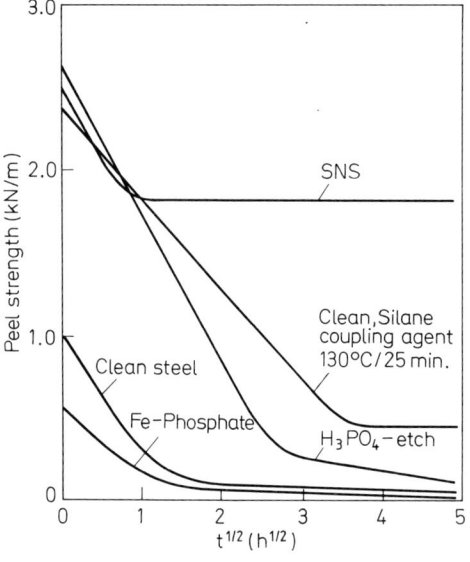

Fig. 13. A comparison of the peel strengths of steel/epoxy coating systems prepared using the SNS pretreatment and various other common pretreatments as a function of immersion time in water at 72 °C [29] (Reprinted from Ref. 29, p. 406, by courtesy of Plenum Press)

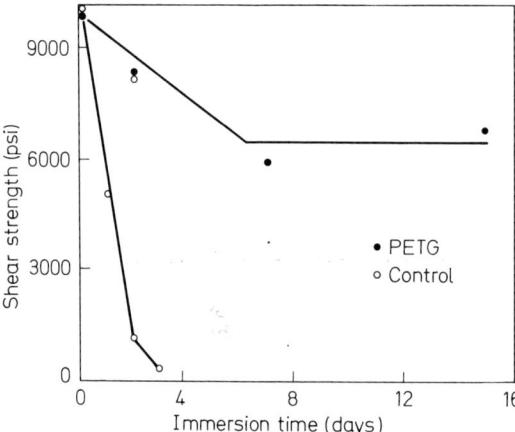

Fig. 14. Effect of pentaerithritol tetrathioglycolate (PETG) coupling agent treatment on the durability of steel/epoxy napkin ring joints [124]

form a five-membered chelate ring with ions on the steel surface and complete the chemical coupling by reacting with epoxide groups in the resin [124].

In addition, by attaching an amino-functional group to a benzotriazole compound which is known to be an effective corrosion-inhibitor for copper, Park and Bell [17] developed a coupling agent which has been shown to significantly increase the durability of copper/epoxy napkin ring joints in water. The joints treated with 5-aminobenzotriazole exhibited a more than 500% better breaking stress than similarly prepared controls after exposure to boiling water for 800 hours. Furthermore, Park [127] found that multifunctional mercaptoester coupling agents also served as very successful coupling agents for copper/epoxy coating systems.

Present day multifunctional chemical coupling agents for metal/epoxy systems provide many reactive sites that can form chemical bonds with both the metal and the epoxy resin. These coupling agents are advantageous because the large number of functional groups increases the probability of a significant number of successful covalent bonds forming across the interface. However, these coupling agents also exhibit some unfavorable properties which limit their effectivness. The functional groups are generally polar in nature and hence the coupling agent region is usually quite hydrophilic. Therefore, although these coupling agents can increase the integrity of the interfacial region, they also can cause an increased infiltration of destructive water molecules into this region. Hence, the strength loss mechanisms that can proceed in the interfacial region in the presence of water can actually be accelerated by the presence of the coupling agent layer. Also, the polar functional groups of chemical coupling agents form bonds with the metal or epoxy which are often susceptible to hydrolysis. The result is that coupling agents will not always provide an increase in the durability of a metal/epoxy adhesion system.

In addition, it is very difficult to achieve a desired monolayer coverage of the coupling agent molecules on the substrate surface. Several layers of physically adsorbed coupling agent molecules often accumulate. Since these layers are usually cohesively weak, the coupling agent region can become the weakest link in the adhesion system. Indeed, it has been shown using Auger [62] and X-ray photoelectron spectroscopy [62,124] that fracture through the coupling agent region often occurs. The increased durability that has been achieved using silanes and mercaptoesters can be explained by the fact that these coupling agents have the capability of self-polymerizing and crosslinking to form network structures which are cohesively strong. However, since mercaptoesters are guite hydrophilic and both silane and mercaptoester coupling agent regions are susceptible to hydrolysis, the increase in durability that can be achieved by using these coupling agents will reach a limit.

On the other hand, new combined coupling agent — corrosion inhibitor pretreatment processes [29,125] have provided a new approach to increasing metal/epoxy adhesion systems with limits that are not yet known.

Despite the many unfavorable properties of present day chemical coupling agents, the authors still believe that they have the potential to be a successful means to significantly increase the durability of metal/epoxy adhesion systems. The authors are currently investigating the feasibility of using polymeric coupling agents with a relatively small number of functional groups in an attempt to considerably increase the toughness and hydrophobicity of the coupling agent region and yet obtain a significant degree of chemical bonding across the metal/epoxy interface.

6.2 Formation of Metal Oxides Which Promote Mechanical Aspects of Adhesion

Recently, many research efforts have been directed at developing pretreatments for metal surfaces which produce oxide layers with pores, fibrous projections, or microscopic roughness which can enhance metal/polymer adhesion by mechanical means. In order for the pretreatments to lead to an increase in durability, the oxide layers formed must be stable under environmental conditions. The bulk [31,33,52,128] of the research in this area has been completed in an attempt to increase the durability of aluminum/polymer joints used in the aerospace industry. However, similar pretreatments for titanium [129, 130], copper [36, 38] and steel [131] have also been investigated. Because studies using aluminum have been most extensive, this discussion will focus primarily on the advances that have been made in increasing the durability of aluminum/epoxy adhesion systems through the use of various pretreatments.

For years, the Forest Products Laboratories (FPL) process [31], which consists in etching the aluminum substrate in an aqueous sodium dichromate acid solution, has been used to prepare aluminum for adhesive bonding. However, it has been discovered [129, 132, 133] that, by applying an anodizing potential to aluminum in the presence of an agressive electrolyte such as phosphoric acid, an oxide layer can be produced which leads to an increase in the initial strength and the long-term durability of aluminum/epoxy adhesion systems.

The FPL and phosphoric acid anodizing (PAA) pretreatments both dissolve away the original oxide layer in the early stages of the pretreatment. For example, an FPL oxide has been reported by Venables [86] to dissolve completely after only 30 seconds in the PAA electrolyte. Therefore, the new oxide layer developed is essentially independent of prior handling or pretreatments. With the aid of a scanning transmission electron microscope, Venables, McNamara, Chen and Sun [52] have proposed structures for the oxide films formed using the FPL (Fig. 15a) and PAA (Fig. 15b) pretreatments. Observing Figs. 15a and 15b, one can see that the oxide produced by the PAA pretreatment is thicker and possesses longer fibrous projections and a more developed cell structure in comparison to the oxide produced by the FPL pretreatment [86]. However, Pocius [33] has shown that the addition of predissolved aluminum alloy to the FPL etch bath can increase the thickness of the oxides formed by this method. On the other hand, the oxides produced by both pretreatments have pores which are large enough (~40 nm diameter) for epoxy resin and curing agent prepolymeric molecules to penetrate into [134]. Upon curing, the epoxy resin becomes mechanically embedded into the oxide structure. Therefore, both the FPL and PAA pretreatments can enhance adhesion by mechanical aspects.

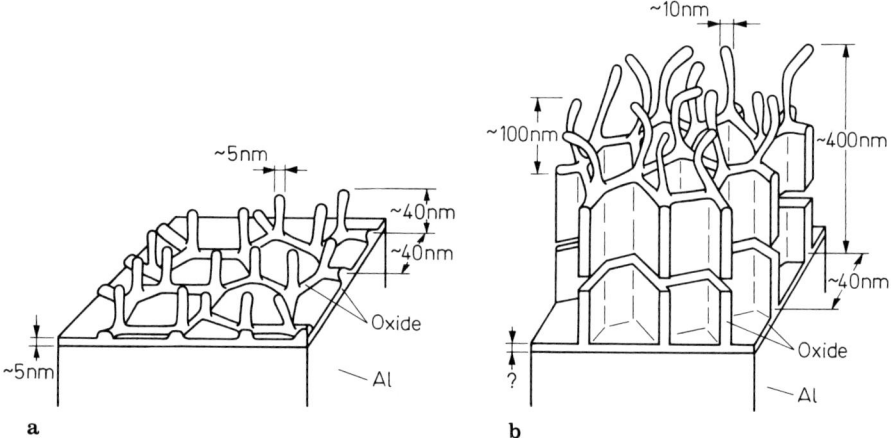

Fig. 15a and b. Perspective of the proposed (Venables et al. [52]) oxide morphology produced on aluminium by the **a)** FPL and **b)** PAA processes [52]. (Reprinted with permission from Chapman and Hall LTD.)

Since water exposure has been shown [86] to have no substantial *short-term* effect on adhesive bonds in which a large degree of mechanical interlocking is present, these pretreatments have the potential to enhance the durability of aluminum/polymer adhesion systems.

One might expect that the more developed porous layer produced by the PAA process would tend to provide a greater number of successful mechanical interlocking sites [54, 86, 135, 136] and initiate a larger degree of plastic deformation in the resin upon failure than the FPL oxide. Test data [129] comparing PAA and FPL pretreated systems have supported this conclusion. Also, the oxides formed by the PAA pretreatment have exhibited better stability in wet environments [54]. Hence, the PAA process has replaced the FPL etch as the method of choice for the pretreatment of aluminum for adhesion systems [132, 133].

The hydration of aluminum oxides, with consequent oxide strength loss, was presented as an adhesion strength loss mechanism in Section 4.2.2. Since the conversion of oxide to hydroxide is known to be largely responsible for the loss of strength of aluminum/polymer bonds in wet environments, it follows that the pretreatments discussed in this section will not result in a significant increase in durability unless this strength loss mechanism is retarded. Aluminium/epoxy joints which were formed after employing the PAA pretreatment have exhibited very good durability in the presence of water. On the other hand, FPL treated substrates have resulted in joints with relatively poor durability. Using XPS and surface behavior diagrams, Davis, Sun, Ahearn and Venables [137] attributed the good durability of the PAA systems to the formation of a thin layer (approx. one monolayer) of $AlPO_4$ which covers the oxide produced and protects it from hydration. Although the nature by which the phosphate layer protects the oxide is not completely understood, it has been proposed that the presence of the phosphate layer adds two steps to the aluminium oxide hydration mechanism. The first step is the hydration of the $AlPO_4$ layer. The second is the dissolution of the hydrated phosphate layer. Once the phosphate layer has dissolved, the oxide hydration mechanism proposed in Section 4.2.2. follows. However, since the second step is very slow, it becomes rate limiting and hence the oxide hydration mechanism is successfully retarded. This suggests that the greater durability achieved with the PAA treatment relative to the FPL treatment is primarily due to the presence of the phosphate layer and not the elaborate oxide structure as was first believed [137].

Aluminum/epoxy joints formed using FPL treated substrates exhibit very good initial adhesion strength. In addition, the FPL process is significantly easier to perform than the anodizing pretreatments. Therefore, if the FPL process could be modified so that the oxide layer produced is protected from hydration, it could once again become the pretreatment of choice for aluminum substrates.

Workers at Martin Marietta [86, 137, 138] have been investigating the feasibility of applying a monolayer of inhibitor to aluminum oxides produced by the FPL process in an attempt to protect the oxide from hydration without interfering with the favorable mechanical interlocking features of the oxide structure. They have reported that applying a monolayer of an amino phosphonic acid (dipping in 3 to 300 ppm aqueous solution) significantly improves the stability of FPL produced oxides. Figure 16a [83] illustrates aluminum/thermoset adhesive joints prepared with FPL/inhibitor treated substrates performed as well as joints pepared from PAA treated

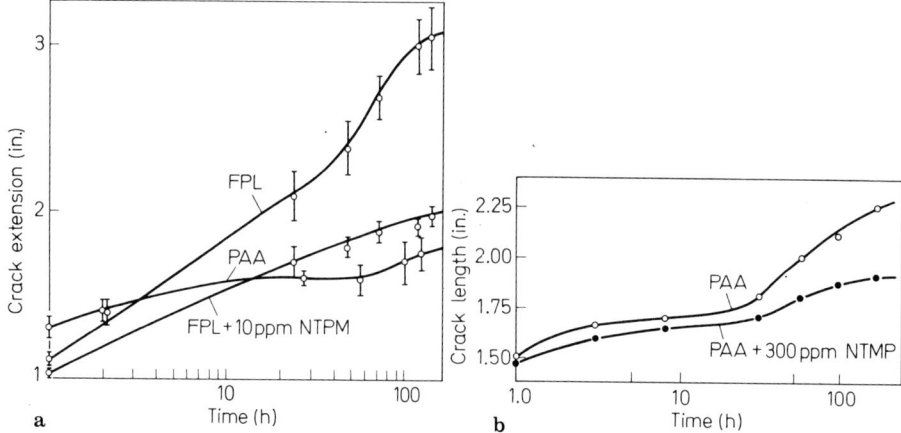

Fig. 16a and b. Wedge-test crack length (in.) of aluminum/thermoset (American Cyanamid FM123-2) joints as a function of exposure time to a 100% r. h., 60 °C environment. **a)** FPL, PAA, FPL + 10 ppm nitrilotris (methylene phosphonic acid (NTMP) [83] and **b)** PAA and PAA + 300 ppm NTMP pretreatments were employed [139] (Reprinted with permission from Chapman and Hall, LTD.)

substrates in wedge tests. Also, Fig. 16b [139] shows that treating PAA treated aluminum with amino phosphonic acid can further increase the stability of the oxide produced, and hence the durability of aluminum/thermoset joints, beyond that achieved with the PAA treatment alone.

Additional work must be completed before these hydration inhibitor treatments will be widely used. However, it appears that combined FPL/inhibitor pretreatments have the potential of producing water stable aluminum oxides with structures that promote mechanical aspects of adhesion in a relatively simple manner. Since 'mechanical adhesion' mechanisms are not greatly affected by water, these pretreatments show promise as a means of increasing the durability of metal/polymer adhesion systems in wet environments.

6.3 Methods to Prevent Corrosion-Induced Delamination

In Section 4.2.3, the concept was introduced that corrosion of the metal substrate can play an important role in the reduction of the adhesion strength of metal/epoxy coating systems, if a defect or discontinuity is present in the coating [5, 87–92]. However, corrosion processes are generally believed not to be important in strength loss mechanisms of metal/epoxy adhesive joint systems [75, 80]. The cathodic reaction involving the formation of hydroxide ions at the metal/epoxy interface has been singled out as the corrosion reaction responsible for the loss of adhesion in coating systems [5, 90, 91]. Five components are required to be present at the metal/epoxy interface for hydroxide formation to occur; namely [91] (1) water, (2) oxygen, (3) electrons, (4) cation counterions, and (5) a catalytically active metal oxide. Therefore, by preventing just one of these five required components from being present in the interfacial region, this destructive cathodic corrosion reaction could be stopped.

Although the fundamentals governing the corrosion processes under protective coatings are generally well understood, little advancement has been made in developing methods to successfully prevent such corrosion. One difficulty is that a large number of different accelerated tests are commonly used to determine how effective protective coatings are in preventing corrosion. Unfortunately, only long-term practical exposure tests have been found to give reliable results [140]. Another problem is that the results from these tests have in large part been a report of the appearance of the samples after exposure. Therefore, the evaluation has been based solely on the final result, and not on what properties of the coating or the metal pretreatment were responsible for the favorable or unfavorable results observed [140].

In this section, possible methods of preventing or minimizing corrosion-induced delamination in metal/epoxy coating systems will be proposed. However, it is important to remember that in order to develop one of the proposed methods to the extent where its full potential can be realized, tests must be completed which examine the science responsible for the good and poor protective properties observed. In addition, the effect of each method on other important properties of the coating system, such as adhesion strength, durability, processability, toughness and cost must be investigated.

6.3.1 Decreased Water Permeation Through the Epoxy Coating

Water can permeate through all organic polymer coatings to some extent. The degree with which water can permeate through epoxy coatings (permeability coefficient, $P = 0$–40×10^{-9} ([ml at S.T.P.]-cm/cm^2 s-cmHg); diffusion constant, $D = 2$ to 8×10^{-9} (cm^2/s) at 25 °C) [1,141] is relatively low in comparison to most polymer coatings. However, Funke and Haagen [140] and others [142,143] have reported that the water permeation rate is sufficiently high so that if the coating thickness and pressure drops are in the practical range and the other necessary components were present, the corrosion reactions could proceed on a steel surface as rapidly as if no coating were present.

One method that has shown promise for reducing the permeation of water through epoxy coatings involves the introduction of fluorine into the epoxy network. Highly fluorinated epoxy resins of the form shown below have been developed by Griffith,

O'Rear, Reines and Bultman [144–146] (R_f represents a perfluorinated alkyl group which can be varied to change the fluorine content of the resin). Fluorinated epoxy coatings which have been pigmented with poly(tetrafluoroethylene) powder can reduce the rate by which water reaches the corrodible metal surface by reducing the solubility of water in the coating. In fact, the amount of water absorbed into fluorinated epoxy resin coatings has been shown to be as much as 85% less than that absorbed by a conventional epoxy coating [146]. Fluorinated epoxy resins, therefore, have the potential to retard some of the water-induced adhesion strength loss mechanisms by reducing the availability of water at the interfacial region. Unfortunately, the water permeation rate through these resins is still sufficiently high for the modification of the resin to be ineffective in reducing the rates of corrosion processes.

6.3.2 Decreased Oxygen Permeation Through the Epoxy Coating

Guruviah [142] and Baumann [143] have compared oxygen permeation rates through polymer coatings with the amount of oxygen required for atmospheric corrosion of bare steel, within a range of practical corrosion rates. Their results indicate that the permeation rate of oxygen through an epoxy resin coating (0.2–6.3×10^{-5} g cm^{-2}/d^{-1} (d = day), thickness 20–35 μm) [140] is within the lower end of the range of values determined for the amount of oxygen consumed (0.87–15.13×10^{-5} g cm^{-2} d^{-1}) [143] by steel corroding under various practical conditions. Therefore, if corrosion does occur under an epoxy based coating, the rate of the corrosion processes may be limited by the rate of oxygen permeation through the coating [140].

Funke and Haagan [140] found that the rate of oxygen permeation through organic coatings could be decreased markedly by adding pigments or fillers to the coating system. Also, they concluded that the effectiveness of different pigments and fillers at reducing oxygen permeation is most likely dependent solely on their physical shape. This explains why the addition of pigments like iron oxide and mica to coating systems have provided enhanced corrosion protection, perhaps by decreasing the cross-sectional area available to the diffusing specie. Therefore, if adhesion strength loss in a metal/epoxy coating system is known to be corrosion-induced, reducing the availability of oxygen in the interfacial region by adding primers or fillers provides a means of retarding this strength loss mechanism.

Leidheiser [91] has proposed numerous methods of attack which can be used in an attempt to prevent or minimize the corrosion-induced delamination of polymer coatings from metal substrates. All of these methods involve the control of the chemical and physical properties of the interfacial layer between the coating and the metal substrate. Three of these methods are discussed in Sections 6.3.3–6.3.5.

6.3.3 Reduced Electrical Conductivity of the Oxide Layer

In corrosion-induced delamination, the electrons required for the cathodic reaction must pass through the metal oxide present at the substrate/coating interface. Usually the oxide film is very thin. Therefore, for most cases, the electrons can pass through the oxide layer under a very small potential gradient [91]. However, Leidheiser [91] has pointed out that if the oxide layer is a poor electrical conductor (i.e. aluminum oxide), and is also relatively thick (>30 Å), the potential gradient will be too large for the electrons produced by the anodic reaction to pass through at a significant rate. The result is that all the electrons flow through the coating defect area where the oxide layer has been destroyed, rather than through the oxide layer under the intact coating. Since no electrons are present under the coating in the area adjacent to the defect corrosion-induced delamination should not occur.

The high electrical resistivity of aluminum oxide is believed to be the major reason why coatings continue to exhibit very strong adhesion to aluminium substrates even when localized corrosion is observed to occur. Therefore, by developing a pretreatment process for any metal substrate which produces a metal oxide with high electrical resistivity, the resulting metal/epoxy coating system should show enhanced resistance to corrosion-induced delamination [91].

6.3.4 Incorporation of Cation-Exchange Materials Into the Metal/Epoxy Interfacial Region

Leidheiser and Wang [92] have shown that the identity of the cation present in the electrolyte can affect the rate at which corrosion-induced delamination occurs in steel/epoxy coating systems. From their observations they conclude that the rate by which cation counterions diffuse through the epoxy coating can limit the rate of hydroxide formation by the primary cathodic reaction. As a result of these observations, Leidheiser [91] proposed the incorporation of a cation-exchange material into the interfacial region of metal/polymer coating systems. The cation-exchange material must adsorb the diffusing cation with the simultaneous release of a hydrogen ion for it to be effective. The released hydrogen ion would prevent the interfacial region from becoming highly alkaline, by reacting with the hydroxide ions formed by the cathodic reaction. Control of the pH would be beneficial because the degradation of the epoxy coating in the interfacial region by hydrolysis, an important step in the corrosion-induced delamination mechanism, requires a highly alkaline environment in order to proceed at a significant rate [91].

A simple and effective way of incorporating such a cation-exchange material into the interfacial region would be to apply it to the metal surface as a final pretreatment process. This way the released hydrogen ions would be present in the immediate vicinity of the hydroxide ion generation sites. Therefore, the hydroxide ions could be promptly neutralized and the hydrolysis of the epoxy coating by strong alkali minimized [91].

6.3.5 Use of Inhibitors

Most corrosion inhibitors are of the adsorption type. In general, these are compounds which adsorb on the metal surface and act to suppress anodic and/or cathodic corrosion processes [147].

Leidheiser and Wang [92] have shown that the use of an inhibitor can reduce the rate of the primary cathodic reaction in the corrosion-induced delamination mechanism by reducing the catalytic activity of the metal oxide. For this case the catalytic activity of zinc oxide on galvanized steel was reduced by a dipping pretreatment in a $CoCl_2$ inhibitor solution. This reduction in activity was confirmed using cathodic polarization data. In addition, Fig. 17 [91] illustrates that the pretreatment retards the cathodic delamination of epoxy coatings from galvanized steel substrates. Leidheiser and Suzuki [148] have attributed the decrease in catalytic activity to the incorporation of cobalt into the surface of the zinc oxide layer. Furthermore, they suggest that the cobalt ions in the oxide surface can successfully 'trap' electrons by the following reaction:

$$Co^{+2} \text{ (in oxide)} + 2e^- = Co^0 \text{ (in oxide)}$$

The proposed result is the presence of an excess of Zn^{+2} in the oxide lattice which can also 'trap' electrons, thus reducing the number available for the cathodic corrosion reactions [148].

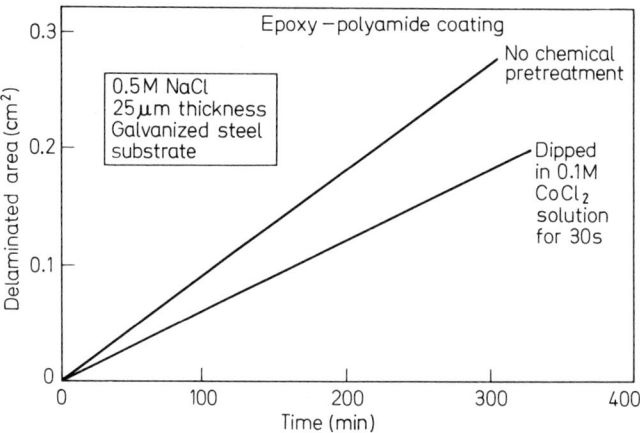

Fig. 17. Cathodic delamination rates of galvanized steel/epoxy-polyamide coating systems when a 0.1 M CoCl₂ dipping pretreatment was applied to the metal substrate prior to the application of the coating [91]. (Reprinted with permission from Ref. [91]. Copyright (1981) American Chemical Society)

A thorough review of inhibitors along with their corrosion retarding mechanisms has been presented elsewhere by Trabanelli and Carassiti [149].

Five methods were discussed as means for reducing the rate of corrosion-induced delamination. The methods which involve decreasing the rate of oxygen permeation through the epoxy coating, increasing the electrical resistivity of the metal oxide layer, and reducing the catalytic activity of the oxide surface are believed by the authors to show the most promise for successfully reducing the rate of the adhesion strength loss mechanism. This is due to the fact that all of these methods involve relatively simple chemical and physical principles which should require, at most, the addition of one simple step in coating system fabrication processes.

6.4 Relieving Internal Stresses

The development of internal stresses in metal/polymer adhesive and coating systems due to shrinkage of the polymer upon setting and to thermal expansion differences was discussed in Section 5. The presence of internal stresses in an adhesive or coating can significantly reduce the adhesion strength of the metal/polymer system. If the internal stress in an adhesive is σ_i, and σ_0 is the fracture strength of a stress-free adhesive bond, then the actual fracture strength of the system, σ, is [150]:

$$\sigma = \sigma_0 - \sigma_i \qquad (6.4\text{-}1)$$

Therefore, relieving internal stresses that are present in polymer adhesives and coatings should result in an increase in the strength of metal/polymer adhesion systems [150].

6.4.1 Addition of Fillers

Adding fillers to epoxy formulations can reduce the magnitude of the internal stresses developed upon setting because they act to reduce the difference between the thermal expansion coefficients of the adhesive and adherend by decreasing the thermal expansion coefficient of the epoxy adhesive or coating. The addition of fillers such as alumina or calcium carbonate have been found to increase the adhesion strength of metal/epoxy systems [9,151]. Fillers are also advantageous because they can reduce the cost and improve the abrasion resistance of the epoxy adhesive or coating. However, not all fillers are beneficial. When choosing a filler for an epoxy formulation, it is important to be sure that it will not interact chemically with the epoxy resin in a manner such that the resin's physical and chemical properties are greatly altered [9,96]. Fillers can also be detrimental to an adhesion system by interfering with the development of adhesion forces in the interfacial region or by introducing sharp edges which can lead to excessive stress concentration.

6.4.2 Addition of Flexibilizers

The build-up of internal stresses can also be reduced by adding a flexibilizer to the epoxy adhesive or coating. A monofunctional flexibilizer for epoxy resins usually consists of a long flexible chain with an epoxy functional group at one end (i.e. epoxidized vegetable oils or similar compounds) [150]. When the curing reaction proceeds primarily through the reaction of epoxy groups, the addition of monofunctional flexibilizers decreases the functionality of the system and introduces long free floating chains into the epoxy network. The presence of these floating chains produces a much more loose network which allows a greater degree of segment mobility. The enhanced mobility increases the ability of the system to relax the internal stresses that are developed upon setting. However, since flexibilizers will also reduce the modulus and glass transition temperature of the epoxy, this option may result in an epoxy coating or adhesive which does not posses the required physical properties for the desired application, and corrosion may increase owing to enhanced diffusion rates [96,150].

The addition of plasticizer to the epoxy resin can produce the same effect as described for the addition of flexibilizers. In fact water, which has been shown to have many detrimental effects on the adhesion strength of metal/epoxy systems, can act as a plasticizer and hence reduce the internal stresses in an epoxy resin adhesion system. However, experience indicates that the many harmful effects of water in regards to metal/epoxy adhesion systems heavily outweight this one possible beneficial attribute. Also, it has been shown by Sargent [152] that although water can act to relieve shrinkage stresses in metal/epoxy joint systems, often the water uptake is sufficient to generate swelling stresses in the system. In principle, swelling stresses are just as detrimental to the epoxy resin as shrinkage stresses [96].

7 Techniques Used to Determine the Locus of Failure

Failure can occur in metal/epoxy adhesion systems in any one or more of a number of different regions. The fracture may propagate through the bulk metal or epoxy, the metal oxide layer, the metal oxide/epoxy or metal/metal oxide interfaces, or through weak boundary layers (WBL's) very near the interfaces. Some workers [78,153] believe that most failures that have been claimed to be interfacial have actually propagated through WBL's. Others [154] disagree. The reason this argument is still unresolved is because it is usually not a simple task to determine conclusively where failure has occurred. Observing both fracture surfaces, with several different investigative techniques, is often required.

Following fabrication of adhesive joint or coating systems, various mechanical tests are usually completed before and after exposure to various environmental conditions. These tests are designed to determine the system adhesion strength and durability. An accurate determination of the locus of failure should also be completed because it indicates which link in the system is the weakest. However, often the locus of failure is simply hypothesized from visual observations. This is unfortunate because in order to successfully increase the adhesion strength of an existing system the strength of the weakest link must be increased. Hence, it is important to accurately identify the weakest link; it is ineffective to increase the cohesive strength of the epoxy resin adhesive in a system in which failure propagates through the metal oxide layer.

In order to accurately determine the locus of failure of adhesion systems, the chemistry of the fracture surfaces must be analyzed using surface-sensitive characterization techniques. Many surface analysis techniques are presently available and each technique is based on an intrinsic property of the surface atoms or molecules. Lee [155], Czanderna [156] and Park [157] have reviewed these techniques. However, they suggest that one be aware that new techniques and applications are continually being introduced.

In this section, four of the more popular surface analysis techniques [158], that are currently being used to determine the locus of failure in adhesion systems, will be discussed briefly. The four techniques are ion scattering spectrometry (ISS), secondary ion mass spectrometry (SIMS), Auger electron spectrometry (AES) and X-ray photoelectron spectrometry (XPS). Some of their operating characteristics have been compared by Baun [159] in Table 3. All four of these techniques are based on atomic properties of the surface atoms and molecules. Although they will not be discussed here, rapid advancements are being made in the development of surface characterization techniques based on the vibrational states of molecules [160]. Reflection infrared spectroscopy [161-163], surface Raman spectroscopy [104-166] and inelastic electron tunneling spectroscopy [167] are techniques with which valuable surface chemistry information can be obtained from the vibrational spectra they produce.

7.1 Ion Scattering Spectrometry (ISS) and Secondary Ion Mass Spectrometry (SIMS)

Both ISS [168] and SIMS [169] utilize an ion probe. Some of the bombarding ions will undergo simple binary collisions with substrate surface atoms. A kinetic energy exchange will take place between an incoming ion and the surface atom it collides with. The magnitude of the energy exchange is dependent on the mass of the surface atom. In ISS, the energy loss of the reflected primary beam is analyzed and related to the identity of the surface atoms. ISS is extremely surface sensitive because the possibility of a simple binary collision occurring inside the bulk of the material and the reflected ion escaping without the loss of any additional energy is minimal. The major setbacks of ISS are its poor signal to noise ratio and the fact that the bombarding ions can physically damage the surface that is being analyzed [13].

Table 3. Comparison of primary elemental surface characterization techniques used to determine the locus of failure in adhesion systems [159]. (Reprinted from Ref. 59, p. 136, by courtesy of Plenum Press)

	Ion scattering spectrometry (ISS)	Secondary ion mass spectrometry (SIMS)	Auger electron spectrometry (AES)	X-ray photo-electron spectrometry (XPS)
Principle	elastic binary collision with surface ion	sputtering of surface atoms by ion beam	ejection of Auger electron upon recombination	ejection of photo-electrons by photon
Probe	~1 to 3 keV ions	~1 to 3 keV ions	~1 to 3 keV electrons	0 to 2 keV photons
Signal	ion current versus energy	ion current versus mass	derivation e' emission versus energy	electron emission versus energy
Applicable elements	$Z \geq 3$	all (if positive and negative SIMS)	$Z \geq 3$	$Z \geq 3$
Surface Sensitivity	high	variable	variable	high
Elemental Profiling	yes	yes	yes, with ion beam	yes, with ion beam
Image-spatial Analysis	yes	yes	yes	no
Spectral shift	possible, but generally no	no	yes	yes
Information on chemical combination	yes, in fine features but generally no	in some cases (fingerprint) spectra)	yes	yes
Quantitative analysis	yes	probably no, maybe with similar standards	yes, in principle but difficult	yes
Influence of operating conditions and matrix	no	yes	yes	yes
Isotopic analysis	yes, in principle but generally no because of resolution limits	yes	no	no
Beam induce surface changes	yes, sputtering damage	yes, sputtering damage	yes, due to sputtering and electron beam heating	no, (except when profiling)

On the other hand, SIMS takes advantage of the destructive nature of the ion probe. Atoms can be knocked free (sputtered) from the surface by the bombarding ions and those that become ionized are analyzed by conventional mass spectrometry [170]. A large number of different kinds of ions can be emitted from the surface. The resolution is also quite good. Thus, although SIMS is not as surface sensitive as ISS, it does provide more detailed information about the surface chemistry. ISS and SIMS, therefore, complement one another. Furthermore, since the ion probe sputters away the surface that is being analyzed, the change in the chemistry of the surface as a function of depth below the surface can be studied by these techniques.

7.2 Auger Electron Spectrometry (AES) and X-ray Photoelectron Spectrometry (XPS)

Similar to the techniques described above, AES and XPS [158,171] are often used as complementary techniques in determining the locus of failure in adhesion systems. In AES, the surface of interest is irradiated with a beam of electrons in the energy range of 1–5 KeV, and the energies of the emitted Auger electrons are analyzed. This technique is usually used to obtain an elemental analysis of the surface. A major problem with AES, when it is used for analysis of adhesion systems, is that most organic adhesives are very unstable under the electron beam due to the localized heating that occurs. However, metal surfaces can be readily analyzed using AES [16].

In XPS, on the other hand, photoelectrons, which are emitted when the sample surface is irradiated with a beam of x-rays, are analyzed. The emitted photoelectrons have discrete binding energies that are dependent on both the identity of the parent element and its chemical environment in the surface. Therefore, both the concentration and the chemical state of an element in the surface can be determined. Two advantages of XPS are that the incident x-ray beam is practically harmless to the surface and it also does not induce charging effects, so that the surface chemistry of adhesives and other insulators can readily be investigated [171].

The electron beam used as a probe in AES can be focused to analyze a very small area on the sample surface (diameter 1–50μ) [62,172]. On the other hand, the spatial resolution that can presently be achieved with XPS is relatively poor since it is very difficult to focus the X-ray beam. Therefore, since AES and XPS techniques exhibit complementary strengths, they are often employed together to achieve an accurate determination of the locus of failure in adhesion systems.

Using the four techniques described above, the chemical states of fracture surfaces can be determined, giving rise to information pertaining to the mechanism and locus of failures in adhesion systems. However, to further understand the failure processes, it is helpful to also examine the fracture surfaces with optical and scanning microscopy techniques [173]. This type of investigation can yield information about the morphology of the fracture region. The presence of voids and flaws due to trapped air bubbles, contaminants and incomplete wetting can also be detected [158,163].

If the instrumentation or the knowledge required to interpret the obtained spectra for the four techniques described above is not available, a combined SEM/energy dispersive x-ray spectrometry (EDX) [174] analysis can be used as an alternative method to obtain information about the chemistry and morphology of fracture surfaces. EDX can provide a quick and easy elemental analysis of 'surfaces'. This technique involves detecting the energies of x-rays which are emitted from a surface when it is irradiated with an electron beam (20 KeV source). The major disadvantage of this technique is that the x-rays analyzed are typically generated from 10,000–20,000 Å below the surface. Its relatively poor surface sensitivity generally prevents EDX from being an adequate technique for conclusively determining the locus of failure. However, it does give qualitative information on the elements present on fracture surfaces which can provide meaningful clues as to where the failure actually occurred [174].

With all the information that can be obtained using modern day spectroscopy and microscopy techniques, the failure locations and mechanisms are still not clear in all adhesion systems. However, with these techniques, the insight required to significantly improve metal/polymer adhesion systems using scientific reasoning is now within reach.

8 Effect of Metal Identity

Almost all metals have oxides which, upon interaction with the atmosphere under normal ambient conditions, form similar hydrated, hydroxyl-rich surfaces (see Section 2.1). Hence, one might expect that the interactions between almost all metals and a particular adhesive or coating would be very similar. Bolger, Hausslein and Movlar [15,16] and others [175,176] have shown that this is not the case. Different metals exhibit large differences in their degrees of surface interaction with water and with organic coatings and adhesives. Bolger [15] has attributed these differences partially to the different distributions of surface hydroxyl groups on various metals. However, the fact that the activity of a surface hydroxyl group can be greatly influenced by the identity of the metal atom to which it is attached is the major reason for the observed differences [15].

The durability of metal/epoxy adhesion systems can vary greatly depending on the metal substrate used [16,176]. This statement has been supported by data presented by Bolger et al. [16] which show that copper/epoxy joints exhibit very poor durability upon exposure to boiling water when compared to the durability obtained with steel and aluminum substrates. The primary strength loss mechanisms for each of these metal/epoxy systems are believed to be different [17,71]. In addition, the tensile and corrosion properties of each of the metals differ significantly [147]. Therefore, varying the metal used in an adhesion system can greatly affect the adhesion strength and durability observed.

It is important to remember that this review presents the major factors which influence metal/epoxy adhesion systems in general. As a result, one must realize that additional factors may have to be considered when investigating a specific metal/epoxy adhesion system.

9 Conclusion

The *initial* dry strength that can be achieved with metal/epoxy and, in general, with the majority of metal/polymer adhesion systems is more than adequate for most bonding and coating uses. On the other hand, these adhesion systems generally exhibit poor durability under atmospheric environmental conditions. Water has been shown to reduce adhesion strength by acting to displace the epoxy from the metal substrate, to weaken the metal oxide by hydration, and as a necessary component for corrosion processes which can initiate the delamination of epoxy coatings. The relative importance of each of these mechanisms on the actual strength loss that occurs when metal/epoxy systems are exposed to water varies from system to system. Nonetheless, either separately or collectively, the water-induced strength loss mechanisms can greatly reduce the adhesion strength of all metal/epoxy systems. It is virtually impossible to keep water from eventually reaching the interfacial region under practical conditions. Therefore, the effectiveness of metal/epoxy joint and coating systems has been severely limited.

Internal stresses are formed in metal/epoxy adhesion systems due to the shrinkage of the coating or adhesive upon setting and the difference between the thermal expansion coefficients of the epoxy and the metal substrate. Although they have been largely ignored in many adhesion strength analyses, the presence of internal stresses can be very detrimental to both the wet and dry adhesion strength of metal/epoxy systems. Theories developed by Shimbo, Ochi and Arai [99] and Croll [96] can be used to estimate the magnitude of the internal stresses present and to predict their effect on adhesion strength values, respectively. More work must be completed in order to properly evaluate these theories. However, they should provide a basis for determining the effect of internal stresses on adhesion systems.

Methods have been proposed which have the potential to improve the durability of metal/epoxy adhesion systems in the presence of water. The goal of these methods is not to increase the initial and dry adhesion strengths, but rather to modify the adhesion system so that a reasonable level of adhesion strength can be maintained upon exposure to a wet environment. The authors believe that of the methods proposed in this review, the use of chemical coupling agents and the development of new metal pretreatment processes which produce porous, stable oxides stand out as possessing the greatest potential to significantly increase metal/epoxy adhesion durability.

Chemical coupling agents have the ability to react with both the metal and the epoxy to form chemical bonds across the interface in metal/epoxy adhesion systems. Until the present time, their effectiveness in increasing durability has been limited either by the low cohesive strength of the coupling agent layer or the poor stability of the interfacial bonds formed. However, by directing efforts at improving the strength and stability of the coupling agent region and the interfacial bonds formed in the presence of water and corrosion products, the authors believe that this durability limit can be significantly increased.

Stable metal oxides with porous structures have been produced on aluminum substrates using a PAA or a combined FPL-etch-inhibitor pretreatment [86]. The durability of aluminum/epoxy adhesion systems is greatly enhanced by the use of these pretreatments because the resulting aluminum oxide resists hydration and also provides mechanical interlocking sites for the epoxy adhesive or coating. One would expect that the development of new pretreatment processes which can produce porous, stable oxides on other metals would be benefical in increasing the durability of other metal/epoxy systems. Therefore, the authors expect a large amount of research activity in this area in the future.

Although the two methods discussed above appear to have the greatest potential for increased durability, it is important to realize that simpler methods such as adding fillers to reduce the oxygen permeability coefficient and internal stress build-up in the epoxy also have merit. Regardless of the durability enhancing method employed, spectroscopy and microscopy techniques should be used to determine the locus of failure of the system in adhesion strength tests. This information can be very valuable in terms of providing insight as to how the durability of the adhesion system can be further increased.

Acknowledgement: Preparation of this review was supported in part by the Adhesives and Sealants Council, Inc. U.S.A., which support is gratefully acknowledged.

10 References

1. Kinloch, A. J.: Adhesion. K. W. Allen, (ed.). Applied Science Publishers, London, 1977
2. Potter, W. G.: Uses of epoxy resins. Chem. Publ. Co., New York, 1976
3. Semerdjiev, S.: Metal to metal adhesive bonding. Business Books, London, 1970
4. Gerhart, H. L.: Ind. Eng. Chem. Prod. Res. Dev. *17* (1), 1 (1978)
5. Dickie, R. A.: In: Adhesion aspects of polymeric coatings. K. L. Mittal, (ed.) Plenum Press, New York, 1982, p. 319
6. Nielson, P. O.: Adhesives Age **25** (4), 42 (1982)
7. Bansal, R. K., Singh, M.: Polymer Sci. USSR *22*, 6, 1334 (1980)
8. Grimes, G. C.: Proceedings of the Applied Symposia, *3*, 157 (1966)
9. Lee, H., Neville, K.: Handbook of epoxy resins. McGraw-Hill, New York, 1967
10. Murayama, T., Bell, J. P.: J. Polym. Sci. A-2, *6*, 417 (1970)
11. Bell, J. P.: J. Appl. Polym. Sci. *14*, 1901 (1970)
12. Fowkes, F. M.: In: Treatise on adhesion and adhesives. Vol. 1, R. L. Patrick, (ed.) Marcel Dekker, New York, 1967, Chapt. 9
13. Parks, G. A.: Chem. Review *65*, 177 (1965)
14. Shafrin, E. G., Zisman, W. A.: NRL Report No. 649b. Naval Research Laboratory, Washington, D.C., 1967
15. Bolger, J. C.: In: Adhesion aspects of polymeric coatings. K. L. Mittal, (ed.) Plenum Press, New York, 1982, p. 3
16. Bolger, J. C., Hausslein, R. W., Movlar, H. E.: Int. Copper Research Assoc., Proj. No. 172, Final Report, p. 2, Amicon Corp., Lexington, MA, 1971
17. Park, J. M., Bell, J. P.: In: Adhesion aspects of polymeric coatings. K. L. Mittal, (ed.) Plenum Press, New York, 1982, p. 205
18. Zettlemoyer, A. C.: Chemistry and physics of interfaces. D. E. Gushes, (ed.) Chapt. XII, ACS, Washington, D.C., 1965
19. Brooks, C. S.: J. Coll. Sci. *13*, 532 (1960)
20. Bolger, J. C.: The chemical composition of metal and oxide surfaces and how these interact with polymeric materials, 30th An. Tech. Conf. PE, Chicago, 1972
21. Minford, J. D.: In: Treatise on adhesion and adhesives. Vol. 5, R. L. Patrick, (ed.) Marcel Dekker Inc., New York, 1981, p. 45
22. Allen, K. W., Alsalim, H. S.: J. Adh. *8*, 183 (1977)
23. Minford, J. D.: J. Appl. Polym. Symp. *32*, 91 (1977)
24. Cotter, J. L.: In: Developments in adhesives. Vol. 1, W. C. Wake, (ed.) Applied Science Publishers, London, 1977, p. 1
25. Derjaguin, B. V.: Disc. Faraday Soc. *18*, 26 (1954)
26. Simons, E. N.: The surface treatment of steel. Pitman, London, 1962
27. Gulbransen, E. A., Copson, T. P.: Disc. Faraday Soc. *28*, 229 (1959)
28. Talbot, S., Bigot, J.: Mem. Scient. Revue Métall. *62*, 261 (1965)
29. Graham, L., Emerson, J. A.: Adhesion aspects of polymeric coatings. K. L. Mittal, (ed.) Plenum Press, New York, 1982, p. 395
30. Gulbransen, E. A., Copson, T. P.: Metals Soc. Conf. *4*, 155 (1959)
31. Eichner, H. W.: Forest Products Laboratory Report 1842, Madison, WI. April 1 1954
32. Bethune, A. W.: SAMPE J. *11*, 4 (1975)
33. Pocius, A. V.: In: Adhesion aspects of polymeric coatings. K. L. Mittal, (ed.) Plenum Press, New York, 1982, p. 173
34. Chessen, N., Curran, V.: Appl. Polym. Symp. *3*, 319 (1966)
35. Thrall, E. W.: Adhesives Age Oct. *1979*, 22
36. Evans, J. R. G., Packham, D. E.: J. Adh. *9*, 267 (1978)

37. Hurd, D. T., Krieble, J. G., Pfeiffer, H. G.: Ind. Eng. Chem. *47*, 2483 (1955)
38. Arrowsmith, D. J.: Trans. Inst. Metal Finishing *48*, 88 (1970)
39. Baker, R. G., Spencer, A. T.: Ind. Eng. Chem. *52*, 1015 (1960)
40. Vazirani, H. N.: J. Adh. *1*, 208 (1969)
41. Park, J. M., Bell, J. P.: In: Adhesive joints. K. L. Mittal, (ed.) Plenum Press, New York, 1984, p. 523
42. Smith, T., Kaelble, D. H.: In: Treatise on adhesion and adhesives. Vol. 5, R. L. Patrick, (ed.) Marcel Dekker Inc., New York, 1981, p. 235
43. Battelle Development Corp., Columbus, OH, U.S. Patent 2,864,732
44. Diggle, J. W., Despic, A. R., Bockris, J. O'M.: J. Electrochem. Soc. *116*, 1503 (1969)
45. Allen, K. W., Alsalim, H. S.: J. Adh. *6*, 229 (1974)
46. Allen, K. W., Alsalim, H. S., Wake, W. C.: J. Adh. *6*, 153 (1974)
47. DeLollis, N. J.: Adhesives for metals, Industrial Press, New York *1970*, 35
48. Adams, R. D., Peppiatt, N. A.: J. Strain Anal. *8*, 52 (1973)
49. Kinloch, A. J.: J. Mat. Sci. *17*, 617 (1982)
50. Williams, D. J.: Polymer science and engineering. Prentice-Hall Inc., New Jersey, 1971
51. deBruyne, N. A.: J. Appl. Chem. *6*, 303 (1956)
52. Venables, J. D., McNamara, D. K., Chen, J. M., Sun, T. S.: Appl. Surf. Sci. *3*, 88 (1979)
53. Evans, J. R. G., Packham, D. E.: J. Adh. *10*, 39 (1979)
54. Bascom, W. D.: Adhesives Age *22* (4), 28 (1979)
55. Packham, D. E.: In: Adhesion aspects of polymeric coatings. K. L. Mittal, (ed.) Plenum Press, New York, 1982, p. 19
56. Gent, A. N., Hamed, G. R.: Fundamentals in adhesion. Presented at C. C. Heritage Memorial Workshop, University of Wisconsin, Madison, Aug. 10–14 1981
57. London, F.: Trans. Faraday Soc. *33*, 8 (1937)
58. Cotter, J. L., Hockney, M. G. D.: Int. Metal. Rev. *19*, 103 (1974)
59. Buck, B. I., Hockney, M. G. D.: In: Aspects of adhesion. Vol. 7, K. W. Allen and D. J. Alner, (eds.) Transcripta Books, London, 1973, p. 242
60. Brockmann, W.: In: Adhesion aspects of polymeric coattings. K. L. Mittal, (ed.) Plenum Press, New York, 1982, p. 265
61. Kinloch, A. J.: J. Adh. *10*, 193 (1979)
62. Gettings, M., Baker, F. S., Kinloch, A. J.: J. Appl. Polym. Sci. *21*, 2375 (1977)
63. Watts, J. F., Castle, J. E.: J. Mat. Sci. *18*, 2987 (1983)
64. Brewis, D. M., Comyn, J., Tegg, J. L.: Int. J. Adhes. Adhes. *1*, 35 (1980)
65. DeNicola, Jr., A. J., Bell, J. P.: In: Adhesion aspects of polymeric coatings. K. L. Mittal, (ed.) Plenum Press, New York, 1982, p. 443
66. Kerr, C., MacDonald, N. C., Orman, S.: J. Appl. Chem. *17*, 62 (1967)
67. Noland, J. S.: Polym. Sci. Tech. *9A*, 413 (1975)
68. Butt, R. I., Cotter, J. L.: J. Adh. *8*, 11 (1976)
69. Kerr, C., MacDonald, N. C., Orman, S.: Brit. Polym. J. *2*, 67 (1970)
70. Kerr C., Orman, S.: Brit. Polym. J. *2*, 97 (1970)
71. Garnish, E. W.: In: Adhesion. Vol. 2, K. W. Allen (ed.) Applied Science Pub., London, 1978, p. 35
72. Ruggeri, R. T., Beck, T. R.: Adhesion aspects of polymer coatings. K. L. Mittal, (ed.) Plenum Press, New York, 1982, p. 329
73. Wake, W. C.: Adhesion and the formulations of adhesives. Appl. Sci. Pub., New York, 1982
74. Dunning, W. J.: In: Adhesion. O. D. Eley, (ed.) Clarendon Press, Oxford, 1961, p. 57
75. Gledhill, R. A., Kinloch, A. J.: J. Adh. *6*, 315 (1974)
76. Reynolds, B. L.: Proceedings of Army Materials Technical Conference, *4*, 605 (1975)
77. Gledhill, R. A., Kinloch, A. J.: Polym. Eng. Sci. *19*, 82 (1979)
78. Bikerman, J. J.: The science of adhesive joints. Academic Press, New York, 1968
79. Delollis, N. J., Montaya, O.: J. Appl. Polym. Sci. *11*, 983 (1967)
80. Kinloch, A. J., Gledhill, R. A., Dukes, W. A.: In: Adhesion science and technology. L. H. Lee, (ed.) Plenum Press, New York, 1975, p. 597
81. Koboyashi, G. S., Donnelly, D. J.: Boeing Aircraft Comp. Rep. DG-41517, 1974
82. Biljmer, P. F. A.: J. Adh. *5*, 319 (1973)

83. Ahearn, J. S., Davis, G. D., Sun, T. S., Venables, J. D.: In: Adhesion aspectes of polymeric coatings. K. L. Mittal, (ed.) Plenum Press, New York, 1982, p. 281
84. Noland, J. S.: In: Adhesion science and technology. L. H. Lee, (ed.) Plenum Press, New York, 1975, p. 413
85. Venables, J. D., McNamara, D. K., Chen, J. M., Ditchek, B. M., Morgenthaler, T. I., Sun, T. S., Hopping, R. L.: Proc. of 12th Nat. SAMPE Technical Conference, Seattle, Oct. *1980*, 909
86. Venables, J. D.: J. Mat. Sci. *19*, 2431 (1984)
87. Smith, A. G., Dickie, R. A.: Ind. Eng. Prod. Res. Dev. *17*, 42 (1978)
88. Leidheiser, Jr. H., Kendig, M. W.: Ind. Eng. Chem. Prod. Res. Dev. *17*, 54 (1978)
89. Dickie, R. A., Smith, A. G.: Chemtech. *10* (1), 31 (1980)
90. Dickie, R. A., Hammond, J. S., Holubka, J. W.: Ind. Eng. Prod. Res. Dev. *20*, 339 (1981)
91. Leidheiser, Jr., H.: Ind. Eng. Chem. Prod. Res. Dev. *20*, 547 (1981)
92. Leidheiser, Jr. H., Wang, W.: J. Coat. Tech. *53*. No. 672, 77 (1981)
93. Ritter, J. J.: J. Coat. Tech. *54*, No. 695 (1982)
94. Hammond, J. S., Holubka, J. W., Dickie, R. A.: J. Coat. Tech. *51*, No. 655, 45 (1979)
95. Malynov, V. D., Nasedkina, A. A., Vainshtein, A. A., Volkov, S. D.: Polym. Mechs. *4* (4), 575 (1968)
96. Croll, S. G.: In: Adhesion aspects of polymeric coatings. K. L. Mittal, (ed.) Plenum Press, New York, 1982, p. 107
97. Croll, S. G.: J. Appl. Polym. Sci. *23*, 847 (1979)
98. Cherry, B. W., Coates, R. H.: In: Aspects of adhesion. Vol. 8, K. W. Allen, (ed.) Transcripta Books, London, 1973, p. 307
99. Shimbo, M., Ochi, M., Arai, K.: J. Coat. Tech. *56*, No. 713, 45 (1984)
100. Delollis, N. J., Montoya, O.: Appl. Polym. Symp. *19*, 417 (1972)
101. Durelli, A. J., Parks, V. J., del Rio, C. J.: Acta. Mech. *3*, 352 (1967)
102. Croll, S. G.: J. Coat. Tech. *53*, No. 672, 85 (1981)
103. Dannenburg, H.: SPE J. *21*, 669 (1965)
104. Bullet, T. R.: J. Adh. *1* (1), 73 (1972)
105. Plueddemann, E. P.: 25th SPI Reinforced Plastics Composites Division Conference, Washington, D.C., 1970
106. Shimbo, M., Ochi, M., Shigeta, M.: J. Appl. Polym. Sci. *26*, 2265 (1981)
107. Saarnak, A., Nilsson, E., Kornum, L. O.: J. Oil. Col. Chem. Assoc. *59* (12), 427 (1976)
108. Prosser, J. L.: Mod. Paint Coat. *67* (7), 47 (1977)
109. Gent, A. N.: Int. J. Adh. and Adhes. April, *1981*, 175
110. Ahagon, A., Gent, A. N.: J. Polym. Sci.: Polym. Phys. Ed. *13*, 1285 (1975)
111. Ahagon, A., Gent, A. N., Hsu, E. C.: Adhesion science and technology. In: Polymer Science and Technology. Vol. 9a, L. H. Lee (ed.) Plenum Press, New York, 1975, 281
112. Gent, A. N.: Adhesives Age **25** (2), 27 (1982)
113. Plueddemann, E. P.: Composite materials. In: Interfaces in Polymer Matrix Composites. Vol. 6, Academic Press, 1974
114. Plueddemann, E. P.: In: Treatise on coatings. Vol. 1, Part III, R. R. Meyers and J. S. Long, (eds.) Marcel Dekker Inc., 1972, p. 381
115. Dukes, W. A., Kinloch, A. J.: Preparation of surfaces for adhesive bonding. Explosive Research and Development Establishment, Watham Abbey, England, 1976
116. Walker, P.: J. Coat. Tech. *52*, No. 670, 49 (1980)
117. Clark, H. A., Plueddemann, E. P.: Mod. Plas. *40* (6), 133 (1963)
118. Kaas, R. L., Kardos, J. L.: SPE, 32nd ANTEC, *1976*, 22
119. Bascom, W. D.: Macro molecular *5*, 792 (1972)
120. Cleveland society for coatings technology: J. Coat. Tech. *51*, No. 655, 38 (1979)
121. Kam, T. T., Hon, R. K.: J. Coat. Tech. *55*, No. 697, 39 (1983)
122. Monte, S. J., Sugarman, G.: In: Additives for plastics. Vol. 1, R. B. Seymour, (ed.) Academic Press, 1978, p. 169
123. Bell, J. P., DeNicola, Jr., A. J.: U.S. Pat. 4,448,847, May 15, 1984
124. DeNicola, Jr., A. J.: Ph.D. Thesis, University of Connecticut, 1981
125. Calvert, P. D., Lalanandham, R. R., Walton, D. R. M.: In: Adhesion aspects of polymeric coatings. K. L. Mittal, (ed.) Plenum Press, New York, 1982, p. 457

126. Lin, C. J., Bell, J. P.: J. Polym. Sci. *16*, 1721 (1972)
127. Park, J. M.: Ph.D. Thesis, University of Connecticut, 1983
128. Bethune, A. W., McMillan, J. C., Donnelly, D. J., Kobayashi, G. S.: Boeing Co. Report DG-41517, 1974
129. Ditchek, B. M., Breen, K. R., Sun, T. S., Venables, J. D.: In: Proceedings of the 12th SAMPE Technical Conference. M. Smith, (ed.) Seattle, Oct. *1980*, p. 882
130. Natan, M., Venables, J. D.: J. Adh. *15*, 125 (1983)
131. Evans, J. R. G., Packham, D. E.: J. Adh. *10*, 177 (1979)
132. Farkas, G.: Surf. *14* (93), 37 (1975)
133. McMillan, J. C., Quinlivan, J. T., Davis, R. A.: SAMPE Quarterly *7* (3), 13 (1976)
134. Malpass, B. W., Packham, D. E., Bright, K.: J. Appl. Polym. Sci. *18*, 3249 (1974)
135. Brockmann, W., Kollek, H.: Proc. 23rd Nat. SAMPE Sympos. *1978*, 1119
136. Gettings, M., Kinloch, A. J.: Surface characterization and adhesive bonding of stainless steels. U. K. Atomic Energy Authority, Harwell, England, 1978
137. Davis, G. D., Sun, T. S., Ahearn, J. S., Venables, J. D.: J. Mat. Sci. *17*, 1807 (1982)
138. Venables, J. D., Tadros, M. E., Ditchek, B. M.: U.S. Pat. 4,308,079
139. Hardwick, D. A., Ahearn, J. S., Venables, J. D.: second-year report to Office of Naval Research, Boston, MA., ONR Contract N00014-00-0718, November, 1982
140. Funke, W., Haagen, H.: Ind. Eng. Chem. Prod. Res. Dev. *17* (1), 50 (1978)
141. Barrie, J. A.: In: Diffusion of polymers. J. Crank and G. S. Park, (eds.) Academic Press, London, 1968, p. 259
142. Guruviah, S.: J. Oil. Colour. Chem. Assoc. *53*, 669 (1970)
143. Baumann, K.: Plaste Kautsch. *19*, 455 (1972)
144. Griffith, J. R., O'Rear, J. G., Reines, S. A.: Chemtech. *2*, 311 (1972)
145. O'Rear, J. G., Griffith, J. R.: Am. Chem. Soc., Div. Org. Coat. Plast., — Chem. Prepr. *33* (1), 657 (1973).
146. Griffith, J. R., Bultman, J. D.: Ind. Eng. Chem. Prod. Res. Dev. 17 (1), 8 (1978)
147. Fontana, M. G., Greene, N. D.: Corrosion engineering. 2nd edition, McGraw Hill, Inc., U.S.A. 1978
148. Leidheiser, Jr., H., Suzuki, I.: J. Electrochem. Soc., *242*, 128 (1981)
149. Trabanelli, G., Carassiti, V.: In: Advances in corrosion science and technology. Vol. 1, Plenum Press, 1970, p. 147
150. Wu, S.: Polymer interface and adhesion. Marcel Dekker, Inc., New York, 1982
151. Wiles, U. S. Pat. 2,528,933 (1950)
152. Sargent, J. P.: In: Adhesive joints. K. L. Mittal, (ed.) Plenum Press, New York, 1984, p. 151
153. Brockmann, W., Hennemann, O.-D., Kollek, H.: In: Adhesive joints. K. L. Mittal, (ed.) Plenum Press, New York, 1984, p. 469
154. Good, R. J.: In: Adhesion measurement of thin films, thick films and bulk coatings. K. L. Mittal, (ed.) ASTMSTP 640, ASTM, Philadelphia, PA., 1978, p. 41
155. Lee, L. H. (ed.): Characterization of metals and polymers. Vol. I, II, Academic Press, New York, 1977
156. Czanderna, A. W. (ed.): Methods of surface analysis. Elsevier, New York, 1975
157. Park, R. L.: In: Surface analysis techniques for metallurgical applications. R. Carbonara and J. Cuthill, (eds.) ASTM, Philadelphia, PA., 1976, p. 3
158. Baun, W. L.: In: Adhesive joints. K. L. Mittal (ed.) Plenum Press, New York, 1984, p. 3
159. Baun, W. L.: In: Adhesion aspects of polymeric coatings. K. L. Mittal, (ed.) Plenum Press, New York, 1982, p. 131
160. Allara, D. A.: In: Industrial applications of surface analysis. L. A. Casper and C. J. Powell (eds.) ACS Symp. Series 199, Washington, D. C., 1982, p. 33
161. Harrick, N. J.: Internal reflection spectroscopy. John Wiley and Sons, New York, 1967
162. Allara, D. A.: In: Vibrational spectroscopies for adsorbed species. Vol. 137, M. L. Hair and A. T. Bell, (eds.) ACS, Washington, D.C., 1980, Chapt. 3
163. Tompkins, H. G.: In: Methods of surface analysis. Vol. 1, A. W. Czanderna, (ed.) Elsevier, New York, 1975, Chapt. 10
164. Jeanmaire, O. L., Van Duyne, R. P.: J. Electroanal. Chem. *84*, 1 (1977)
165. Schlotter, N. E., Rabolt, J. F.: J. Phys. Chem. *88* (10), 2062 (1984)
166. Schlotter, N. E., Rabolt, J. F.: Appl. Spectro. *38* (2), 208 (1984)

167. Khanna, S. K., Lambe, J.: Science *1983*, 1345
168. Smith, D. P.: J. Appl. Phys. *38*, 340 (1967)
169. Benninghoven, A.: Surf. Sci. *28*, 541 (1971)
170. Johnstone, R. A. W.: Mass spectrometry for organic chemists. Cambridge University Press, London, 1972
171. Carlson, T. A.: Photoelectron and Auger spectroscopy. Plenum Press, New York, 1975
172. MacDonald, N. C., Waldrop, J. R.: Appl. Phys. Lett. *19* (9), 315 (1971)
173. Buchanan, R.: Am. Laboratory April *1983*, 56
174. Beaman, D. R., File, D. M.: Anal. Chem. *48*, 101110 (1976)
175. Black, J. M., Blomquist, R. F.: Mod. Plast. *35* (10), 225 (1956)
176. Black, J. M., Blomquist, R. F.: Ind. Eng. Chem. June *1958*, 48

Editor: K. Dušek
Received March 11, 1985

Application of FT-IR and NMR to Epoxy Resins

Elaine Mertzel and Jack L. Koenig
Department of Macromolecular Science,
Case Western Reserve University,
Cleveland, Ohio 44106/U.S.A.

The application of FT-IR and high resolution solid state NMR to the structural characterization of epoxies is described. Theoretical and experimental background is given and progress to date in these two fields summarized.

1 Introduction 74

2 Basic of FT-IR Spectroscopy 74

3 Sampling Techniques for FT-IR 76

4 The Interpretation of the Infrared Spectra of Epoxies 79

5 FT-IR Studies Composition of Epoxy Resins 86

6 FT-IR Studies of Curing Kinetics and Mechanism of Epoxy Resins 91

7 FT-IR Studies of the Degradation Processes in Epoxy Resins 92

8 Chemical Reactions of Epoxies with Coupling Agents 92

9 Basis of NMR Spectroscopy 93
   9.1 Dipolar Decoupling 93
   9.2 Magic Angle Spinning 93
   9.3 Cross Polarization 94
   9.4 Solution Multinuclear NMR 94

10 Resolution, Molecular Motion and Characterization of Epoxy in
    Solid State NMR 97
   10.1 Molecular Motion in Solid State Proton NMR 97
   10.2 Resolution and Mobility Studies in Solid State C-13 NMR 100

11 References 110

1 Introduction

The analysis of epoxy resins has been a particular challenge for the polymer chemist because of the complexity of the repeating units. The multitude of comonomers, the number and type of initiators, the variety of possible polymerization reactions, the insoluble nature of the product and the susceptibility of the network to hydrolysis and other types of chemical attack. Consequently there has been little knowledge of the structural basis of the physical, chemical and ultimate mechanical properties of the epoxy resins. However, it is essential that knowledge of the structures and curing processes be obtained in order to optimize the performance of the epoxy resins.

This article will review the impact of two powerful new techniques for characterizing epoxy resins at the molecular level — Fourier transform infrared spectroscopy (FT-IR) and high resolution nuclear magnetic resonance (NMR) of solids. Fortunately, these two techniques are not inhibited appreciably by the insoluble nature of the cured resin. Consequently, substantial structural information at the molecular level can be obtained. In this article, the basis of the methods will be briefly described in order to appreciate the nature of the methods followed by a description of the work on epoxies to date and finally some indication will be given of the anticipated contributions of these methods in the future.

2 Basis of FT-IR Spectroscopy

A number of sources describe the application of the interferometer to infrared instrumentation [1-3]. For our purposes, it is sufficient to recognize the basic differences between FT-IR and classical dispersive infrared spectroscopy. A dispersive instrument utilizes a prism or grating to geometrically disperse the infrared radiation. Using a scanning mechanism, the dispersed radiation is passed over a slit system which isolates a narrow frequency range falling on the detector. In this manner, by using a scanning mechanism, the spectrum, that is, the energy transmitted through a sample as a function of frequency, is obtained. The dispersive method is limited in signal-to-noise ratio because most of the source energy is not utilized, that is, it does not fall on the open slits. Using dispersive techniques, 90% of the energy from the source is lost because of the optics of the system. Using an interferometer and Fourier transform techniques, all of the source energy is being utilized during all of the measurement time. In this manner, the signal-to-noise ratio and thus the sensitivity of FT-IR is much higher ($\times 100$) compared to dispersive infrared spectroscopy.

FT-IR utilizes the Michelson interferometer rather than the grating or prism of the dispersive system. The Michelson interferometer has two mutually perpendicular arms. One arm of the interferometer contains a stationary, plane mirror; the other arm contains a moveable mirror. Bisecting the two arms is a beamsplitter which splits the source beam into two equal beams. These two light beams travel their respective paths in the arms of the interferometer and are reflected back to the beam splitter and on to the detector. The two reunited beams will interfere constructively or destructively, depending on their path differences and the wavelengths of the light. When the path lengths in the two arms are the same, all of the frequencies constructively interfere and add coherently to produce a maximum flux at the detector. This initial high flux is termed the center burst.

When the movable mirror is moved into a position producing a difference in the path leghths of the two arms, the frequencies or the light will interfere in a fashion determined by their respective wavelengths with all of them initially exhibiting some destructive interference causing a dramatic fall off of the intensity. As the path length difference increases, some of the shorter wavelengths will start to constructively interfere and the longer wavelenghts will destructively interfere. The flux at the detector is the sum of the energies from all of the wavelenghts making up the incoming light. The detector sees an interferogram with contributions from all of the frequencies. The interferogram is obtained in the time domain (i.e. as a function of the rate of the mirror drive) while the analyst is interested in the spectrum in the frequency domain. The interferogram is transformed into the spectrum by performing a Fourier transformation of the data. The methods of carrying out this transformation as applied to FT-IR have been described in detail [4,5].

In the early days, this Fourier transformation was a time-consuming, expensive and difficult task due to limited computer speed and capacity. However, with the advent of the fast Fourier transform algorithm of Cooley and Tukey [6] and the improvement in computers, this problem has been resolved so that real time spectra can be obtained with the transformation time of the order of fractions of seconds.

The advantages of FT-IR over grating infrared arise from several sources. The Fellget or multiplex advantage arises from the observation of all of the frequencies all of the time. For measurements taken at equal resolution with the same optical throughput and for equal measurement time with the same detector, the signal-to-noise ratio of the spectra from an FT-IR will be $M^{1/2}$ times greater than on a grating instrument where M is the number of resolution elements being examined during the measurement. For the normal range of the mid-infrared, the signal-to-noise improvement would be 60. This multiplex factor alone allows a substantial improvement in the determination of the degree of cure and composition of epoxy resins [7,8].

This multiplex advantage in FT-IR is easiest to understand with respect to the speed of the measurement. For a resolution of 1 cm^{-1} and frequency range of 400 to 4,000, the FT-IR measurement time is 3,600 times faster than with a dispersive instrument. Consequently, when the time of measurement is limited, the FT-IR has an advantage of two orders of magnitude in speed. For studies of the curing of epoxies this time advantage can be extremely valuable [9].

When the time of measurement is not limited, the rapidity of the scanning can be used to signal average the interferograms with an improvement in the signal-to-noise ratio proportional to $N^{1/2}$ where N is the number of scans. Since, in most instances, the scan rate is about one per second, an order of magnitude increase in the signal-to-noise ratio can be obtained with little additonal effort. One can expect an increase in sensitivity of approximately 1% for an order of magnitude increase in the signal-to-noise ratio, depending on the extinction coefficient of the analytical band being measured. The maximum number of scans which can be added to improve the signal-to-noise ratio is determined by the stability of the instrument since ultimately instabilities undermine the effectiveness of continued signal averaging. In our laboratory, a maximum of 30,000 scans have been recorded on a single sample.

The throughput or Jacquinot advantage arises from the observation in the FT-IR that there are no slits and the energy throughput is much higher. The throughput in an FT-IR instrument is determined by the size of the mirrors used in the interferometer. A comparison of commercially available instruments suggests the throughput advantage approaches a factor of 60. For sampling of polymeric films, a larger sample size can be utilized with FT-IR making the measurable signal-to-noise higher and the sensitivity greater. In our laboratory, we have enlarged the aperature of the commercially available liquid cells (designed originally for dispersive instruments) in order to benefit from the greater throughput possible with the FT-IR instrumentation.

The Conne or frequency advantage of FT-IR compared to dispersive instrumentation is that the frequencies of an FT-IR instrument are constantly exhibiting no drift of the type exhibited by dispersive instruments because the frequencies are internally calibrated by a helium-neon laser in an FT-IR instrument. This frequency stability and reproducibility is particularly important for signal averaging and absorbance subtraction of spectra. For the absorbance subtraction technique to be useful for epoxy resins examined over a period of time, such as months or years, long term frequency accuracy must be maintained. Application such as quality control and long term aging and weathering require the reproducibility of the frequency that is achieved with FT-IR instrumentation.

Of course, FT-IR has some inherent disadvantages. One of the most important is that the raw data, an interferogram, is for all practical purposes unintelligible to the analyst's eye. So a computer must be utilized to translate the raw data into interpretable data. In the process, a number of factors such as apodization and phasing are introduced which modify the raw data. The analyst must be knowledgeable about these modifications of his data and be prepared to control these modifications in a reproducible and optimum manner.

3 Sampling Techniques for FT-IR

All of the usual sampling techniques used in infrared spectroscopy can be used with FT-IR instrumentation. The optics of the sampling chamber of commercial FT-IR instruments are the same as the traditional dispersive instruments so the accessories can be used without modification for the most part. To make full use of the larger aperature of the FT-IR instrument, some accessories should be modified to accomodate the larger beam. The instrumental advantages of FT-IR allow one to use a number of sampling techniques which are not effective using dispersive instrumentation. Transmission, diffuse reflectance and internal reflectance techniques are most often used in the study of epoxy resins.

For transmission measurements on epoxy liquids or solids, it is desirable to make full use of the larger beam diameter by using larger sample diameters. Coupled with signal averaging and the other FT-IR advantages, very high signal-to-noise ratios can be obtained. When KBr pellets are used, it is desirable to keep the amount of epoxy at a sufficiently low level so that the absorbance is below 0.80 units. This may seem to be a contradiction since one would perceive, the more sample the greater the signal. However, most FT-IR studies utilize the computer to process the data whether it be for spectral subtraction, curve fitting, deconvolution etc., [10] and it is absolutely necessary to keep the absorbance in the linear range of the Beer-Lambert law or spectral artifacts will be generated during the data processing steps.

On the other hand, if no data processing is involved and one is seeking a very weak signal, the absorbance scale can be expanded to 1/3 the signal-to-noise ratio rather than the highest signal. Of course, a linear optical signal requires that the infrared sample have no residual orientation, no voids or holes, and a uniform distribution of material. Careful control of the sample preparation procedure must be achieved in order that reproducible samples are obtained for the infrared examination.

With FT-IR it should be remembered that the most desirable sample is one producing the highest signal-to-noise ratio rather than the highest signal. Of course, a linear optical signal requires that the infrared sample have no residual orientation, no voids or holes, and a uniform distribution of material. Careful control of the sample preparation procedure must be achieved in order that reproducible samples are obtained for the infrared examination.

Epoxy resins are often a part of a glass fiber reinforced composite. Since glass is a highly polar substance, it exhibits a very strong interfering absorbance spectrum. There are two general approaches to this problem. One is to use the near infrared frequency region where the absorbance of glass is much weaker and better separated from the glass [11]. Another is to separate the glass from the epoxy. Perkinson has developed a differential floatation method to separate the materials [12].

In internal reflection spectroscopy (IRS), the spectrum is obtained with the sample in optical contact with another material (e.g. a prism) and the beam is passed through the prism onto the sample. The prism is optically denser than the sample, the incoming light forms a standing wave pattern at the interface within the dense prism medium whereas in the sample (with the lower refractive index), the amplitude of the electric field falls off exponentially with the distance from the phase boundary. When the sample exhibits absorbance, the reflectance measured is given by:

$$R = 1 - kd_c$$

where R is the measured reflectance, k is the absorptivity and d_c is the effective layer thickness. When the spectra are obtained in this fashion, the technique is often referred to as attenuated total reflection (ATR). When multiple reflections are used to increase the sensitivity, the technique is termed multiple internal reflection (MIR). At first glance, the IRS spectrum and the transmission spectrum of the same sample are very similar. However, closer inspection reveals that the long wavelength side of the absorption bands are distorted in IRS relative to transmission and secondly, the longer wavelengths appear relatively stronger. The distortion of the IRS spectrum is a function of the differences in the refractive indices of the optical element and the sample, the angle of reflection, and the optical constants of the sample. A computer algorithm has been written which utilizes a knowledge of the optical constants and their dispersion to convert an IRS spectrum to its equivalent transmission spectrum [14].

Because, to date, no IRS accessories have been designed to benefit from the larger beam diameter of FT-IR, the spectral improvement achieved with FT-IR IRS is not as great as observed with FT-IR transmission compared to dispersive instruments. However, the signal averaging capability and speed make FT-IR a very useful tool.

For epoxy resins in composites or in other fabricated form, IRS is the only approach which does not require substantial sample modification such as grinding.

Another sampling technique of importance is diffuse reflectance spectroscopy which has come to be termed (DRIFT) (Diffuse Reflectance In Fourier Transform) [13]. The technique has been used in the visible and ultraviolet frequency region for a long time, but it was only recently utilized in the mid-infrared region. DRIFT relies on the measurement of the reflected light from a powdered sample radiated with light. Since some of the light is absorbed and the remainder is reflected, study of the diffuse reflected light can be used to measure the amount absorbed. The requirement is that sufficient scattering occur so that the reflected light is isotropic. With powdered samples, this is possible particularly if diluted with a nonabsorbing powder such as KBr. The Kubelka-Munk theory is used to relate the scattering function $f(R_{00})$ to the absorption coefficient (k) and the scattering coefficient (s):

$$f(R_{00}) = (1 - R_{00})^2 / 2R_{00} = k/s$$

where R_{00} is the absolute reflectance of an infinitely thick layer. In practice, a standard is used and the following ratio is calculated:

$$R'' = R'(\text{sample})/R'(\text{standard})$$

where finely ground KBr has been recommended as the standard [12]. For powdered samples, diffuse reflectance offers considerable advantage particularly since no sample preparation is required. DRIFT can also be used successfully for the study of surfaces and of bulk samples which have been ground.

Another technique which has great potential but has not been used extensively to date with epoxy resins is the photoacoustic technique. The photoacoustic technique is simply the generation of an acoustic signal by a sample exposed to modulated light. The solid sample is placed in an enclosed chamber with a coupling gas such as air, helium or argon and is exposed to modulated light. The sample is heated to the extent that it absorbs the incident light and the energy gained is lost to heat through non-radiative processes. Because the light is modulated, the temperature rise is periodic at the modulation frequency, and it is this periodic temperature rise (typically <0.001 °C) at the surface of the sample that, in turn causes a modulation of the gas pressure in the enclosed chamber. This pressure modulation is an acoustic signal and is detected by a sensitive microphone coupled with the chamber. The photoacoustic technique rises because a gas medium surrounding the sample acts as a piston conducting the acoustic signal from the solid sample to the microphone. The photoacoustic technique has the advantage that no sample preparation is required and samples do not have to be made light transmitting in order to be examined. The photoacoustic technique does not have the inherently high signal-to-noise ratio of the other FT-IR sampling techniques but the ease of sampling makes it a most desirable technique to consider [15]. The photoacoustic technique would be particularly useful for studying the photo, thermal, and environmental aging processes of epoxy composites.

4 The Interpretation of the Infrared Spectra of Epoxies

Although FT-IR allows one to obtain quality spectra of epoxy resin systems, the interpretation aspect of the spectra has not changed. It is still necessary to have some insight into the structural origin of the numerous infrared bands in order to use them

Table 1a. Infrared and Raman Band Assignments for EPON 828 [17]

IR	Raman	Assignment	IR	Raman	Assignment
~3500		ν(OH)	1120 w		
	3210 w			1113 s	
	3151 w		1108		
3123 vw	3125 w		1086	1087	δ(φ-H) in-plane
3098 w			1076 sh		
	3067 s	ν(φ-H)		1055 w	
3057		ν$_{as}$(CH$_2$) epoxy	1036	1040 w	ν$_s$(φ-O—C)
3038		ν(φ-H)	1012	1015 w	δ(φ-H) in-plane
	3006	ν(φ-H)		992 w	
2998 sh		ν$_s$(CH$_2$) epoxy	971		
2968	2969	ν$_{as}$(CH$_3$) + ν$_{as}$(OCH$_2$) ?	936 w	938	δ(φ-H) out-of-plane
2929	2927	ν(CH) epoxy?	916	917	epoxy ring
	2910		~906 sh		
~2890 sh			863	863	epoxy ring
2874	2871	ν$_s$(CH$_3$)		836	
2836	2833 w	ν$_s$(OCH$_2$)	831		δ(φ-H) out-of-plane
2805 vw				824 s	δ(φ) out-of-plane
2756 vw	2755 w		808 sh	809	
	2710 w		772		
~2064 w		φ, disubstituted		765	
1891 w		φ, disubstituted	758		
~1766 w		φ, disubstituted	737 w	737	
1608	1607 s	ν(C=C) φ	727 w		
1583	1583	ν(C=C) φ		677 sh	
1511 s	1510 w	ν(C=C) φ		667	
	1481 w	δ(CH$_2$) epoxy		654	
1470 sh				650	
1457	1462	ν(C=C) φ + δ$_{as}$(CH$_3$)	639 w		
1431 w	1429	δ(OCH$_2$)?	604 w		
1414 w	1414 w	δ(CH) epoxy?	586 w		
1385	1385 vw	δ$_s$(CH$_3$) gem-dimethyl	575	580 w	
1363	1360 vw	δ$_s$(CH$_3$) gem-dimethyl	557		
1347	1348	δ(CH) epoxy	504 w	498 w	
1312 sh	1310 sh		450	455 w	
1298	1298	ν(C—O) + ν(C—C)?		395	δ(φ) out-of-plane
	1261	epoxy ring?		374 sh	
	1254	epoxy ring?		350 w	
1248 s		ν(φ-O)		318 w	
1230 sh	1231	δ(φ-H) in-plane		277 sh	
1185	1187	δ(φ-H) in-plane		240	
1157 w	1154	φ		213 sh	
1133	1134			175 w	

s = strong, sh = shoulder,
w = weak, ν = stretching,
vw = very weak, δ = deformation

Table 1b. Infrared and Raman Band Assignments for Nadic Methyl Anhydride

IR	Raman	Assignment	IR	Raman	Assignment
	3077		1039 w		
3072 vw			1015 w	1017 w	
3058 w	3060		1003 vw	1006 w	
3018 sh		$v(=C-H)$	990 vw	989	
	2993			970 w	
2981	2985 s	$v_{as}(CH_3)$	943 s		anhydride ring
2945		$v_{as}(CH_2)$	929	934 s	anhydride ring
	2924	$v(CH)$ tertiary?	916 s	920	anhydride ring
2917	2916	$v_s(CH_3)$	899 s	902	anhydride ring
2879	2882	$v_s(CH_2)$	868 w	870 w	anhydride ring
2857 sh	2857		853 vw		
2828 vw			843	844 w	
	2744 w		816 sh	819	
1858 s	1859	$v_s(C=O)$	798	799 w	$\delta(=CH)$ out-of-plane
	1852		786 sh		
~1820?	1834?		764 sh		
1780 vs	1781	$v_{as}(C=O)$	758 w	759 w	
1704 w			736 w		
1626 w	1627 s	$v(C=C)$	712	717 w	
	1576		696 w		
1465 w	1465 sh	$\delta(CH_2)$	673 sh	672	
1445	1449	$\delta_{as}(CH_3)$	649 sh		
1382 w	1381	$\delta_s(CH_3)$	635 w		
1345 vw		$\delta(CH)$	622	622 s	
1326 w	1329 w	$\delta(CH)$ in-plane		599	
1312 vw			589 w		
1299			573 w	576 w	
1289	1290 w		536 w	539 w	
1277 vw			506 vw		
1267 w			495 w	498 w	
1256 vw			461 w	465 w	
	1246 w		447 w		
~1240			431 w	434 w	
1228 s	1230 w	$v(C-O)$		427 w	
1216			419 vw		
1194 w			408 vw	412 sh	
1180 vw				376 w	
1137 w	1141			360 w	
1124 vw	1125			330 w	
1106 w	1107			242 w	
1083 s		anhydride ring		211 w	
	1074 w			188 w	
1053 w	1054 w			154 w	

vs = very strong;   sh = shoulder;
s = strong;   v = stretching;
w = weak;   δ = deformation
vw = very weak;

Table 1c. Infrared Special Changes During the Crosslinking of a Stoichiometric Nadic Methyl Anhydride/EPON 828 Mixture Catalyzed by 2.0% Weight Benzyldimethylamine

Intensity Decrease	Assignment	Intensity Increase	Assignment
~3060	$\nu_{as}(CH_2)$ epoxy		
3010	$\nu_s(CH_2)$ epoxy		
		2963	$\nu_{as}(CH_2)$
		~2904	$\nu(CH)$
		2863	$\nu_s(CH_2)$
1858	$\nu_s(C=O)$ anhydride		
1780	$\nu_{as}(C=O)$ anhydride		
		1743	$\nu(C=O)$ ester
		1454	$\delta(CH_2)$
		1398	$\omega(CH_2)$
		1361	
		1332	
		1267	$\nu(C-O) + (C-C)$
1228	$\nu(C-O)$ anhydride		
		1178	$\nu(C-O)$ ester
		1155	$\nu(C-O)$ ester
		1127	$\nu(C-C)?$ ester
		1112	
1083	anhydride ring		
		1056	
		1012	
942	anhydride		
928	anhydride		
915	anhydride + epoxy ring		
898	anhydride		
865	anhydride + epoxy ring?		
842			
798			
713			

ν = stretching;
δ = deformation;
ω = wagging

effectively. Unfortunately, the complexity of the molecules involved make complete band assignments impossible and only superficial assignments are available [16]. Some additional assignments have been made for the diglycidyl ether of biphenol A (DGEBA) and nadic methyl hydride (NMA) and these assignments are given in Table 1 [17].

Since the dimensions of the monomer unit are so large these band assignments would be expected to apply to the cured resins, with the exception of those bands associated with functional groups involved in the curing process. Additional bands will be observed arising from the new structures generated during the cure. In Table 2. the bands observed for three different epoxy resins and the assignments of some of the bands to the various functional groups are listed [18]. In addition, the changes in the various absorptions occuring during degradation are indicated. In this table an (×) indicates

Table 2. Application of Fourier Transform IR to Dehydration Studies of Epoxy Systems. Tentative Infrared Absorption Assignments for the Three Cured Epoxy Resins and the Absorption Variations During Degradation [18]

Wavenumber (cm^{-1})	DGEBA IR[a]	DGEBA TD[b]	DGEBA TO[c]	DGEBA PO[d]	DGEPP IR	DGEPP TD	DGEPP TO	DGEPP PO	DGEBF IR	DGEBF TD	DGEBF TO	DGEBF PO	Functional group	Vibration mode
3570	x[e]								x				R—OH	ν(O—H)
3550		+[f]	+		x	+	+				+		ArOH	ν(O—H)
3525						+	+						ArOOH (O)	ν(O—H)
3430														
3350				+				+				+	R—OOH	ν(O—H)
3300														
3200												+	Ar—OOH	ν(O—H)
3068														
3060					x	−	−	−						
3052	x	−[g]	−	−					x	−	−	−	Arylene	ν(C—H)
3038					x									
3034	x	−	−	−					x	−	−	−	Methyl	ν(C—H)
2970	x	−	−	−		−	−	−						
2935														
2933					x	−	−	−	x	−	−	−	Methylene	ν(C—H)
2930					x	−	−	−						
2880									x					
2876	x	−	−	−										
1808		+	+	+		+	+	+		+	+	+	RC—O—CR (O,O) [h]	ν(C=O)
1790							+							
1784			+	+							+	+	ROOOH (O)	ν(C=O)
1782														
1775					x	−	−	−					Lactone	ν(C=O)

Application of FT-IR and NMR to Epoxy Resins

Column groups: ν(C=O) | Quadrant stretching | Semicircle stretching

Column headers (left to right):
1. RCOOR Lactone
2. RCOR
3. RCHO
4. (o-isopropyl benzoic acid / related structure)
5. RCOOH
6. Phenoxy Phenylene
7. Phenylene
8. Phenylene (mono-sub)
9. Phenylene (para-disub)
10. Phenylene (ortho-sub)
11. Phenylene (meta/para)
12. Phenylene (para)
13. Phenylene (mono)

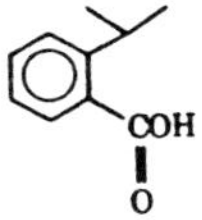

Wavenumber (cm⁻¹)	RCOOR Lactone	RCOR	RCHO	Ar-COOH(i-Pr)	RCOOH	Phenoxy Phenylene	Phenylene	Phenylene	Phenylene	Phenylene	Phenylene	Phenylene	Phenylene
1765	+	+	+		+	+	−	−	−	+		−	+	−
1754	+	+	+		+	+	−	−	−		+	−	+	−
1745	+				+	−	−	−				−	+	−
1732	×		×		×	×	×	×		×			×
1725				+	+	+	−	−	−	+	−	+	−
1715		−		+	+	+	−	−	−	+	−	+	−
1665	+	−		+	+	+	−	−	−		−	+	−
1610		×			×	×	× ×		×				×
1578	+	+	+		+	+	−	−	−	+			−
1510	+	+	+		+	+	−	−	−	+		+	−
1505	+	+	+		+	+	−	−	−	+		+	−
1480	×		×		×	×	× ×					×
1465													
1448													
1440													
1425													
1412													

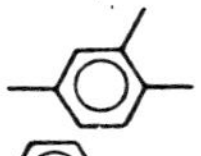

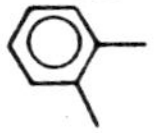

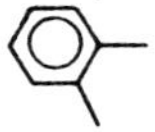

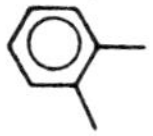

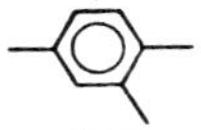

Table 2 (continued)

Wavenumber (cm^{-1})	DGEBA				DGEPP				DGEBF				Functional group	Vibration mode
	IR[a]	TD[b]	TO[c]	PO[d]	IR	TD	TO	PO	IR	TD	TO	PO		
1288					×	−	−	−					Lactone	ν(C—O—C)
1286						+	−	+					Ester or acid	ν(C—O—C) or ν(C—O)
1280			+	+				+		+	+	+	Ar—OH	ν(C—O)
1280			+	+				+		+	+	+	Ar—OH	ν(C—O)
1255	×	−	−	−	×			−				−	Ar—O—R	ν(C—O—C)
1245				+				+				+	Ar—OH	ν(C—O—C)
1202			−	−	×	−	−	−				−	Ar—C—Ar	ν(C—C)
1184	×				×								Ar—C—Ar	ν(C—C)
1180													Ar—C—Ar	ν(C—C)
1170				+	×	+		+		+	+	+	Ar—C—Ar	ν(C—C)
1120	×	−	−	−	×	−	−	−	×			−	Aliphatic chain	ν(C—C)
1096					× ×	− −	− −	− −	×				Aliphatic ether	ν(C—O—C)
1086					×				×					
1035	×				×	−	−	−	×					
1015	×				× × ×	− − −	− − −	− − −					Lactone	ν(C—O—C)
1010	×	−	−	−										
965														
940														
925								+						
890			+										Peroxide	ν(O—O)
885		+		+							+	+		

Wavenumber	IR[a]	TD[b]	TO[c]	PO[d]			Assignment
840	×	−	−	−	+	−	p-Phenylene In-phase, out-of-plane hydrogen wagging
830	×						
825	×		−		−	−	
822							
775	×		+ / −	+ / −	+	−	o-Phenylene In-phase, out-of-plane hydrogen wagging
754	×						
746			+ / −	+ / −	+ −		
740			+	+	+ / −		
736					+	−	o-Phenylene Out-of-plane sextant ring bending
725							
715							
692	×	−	−	−	+	−	

[a] IR: Original IR spectrum of epoxy resin.
[b] TD: Thermal degradation.
[c] TO: Thermooxidative degradation.
[d] PO: Photooxidative degradation; stretching.
[e] ×: Absorption present in the original spectrum.
[f] +: Absorbance increase during degradation.
[g] −: Absorbance decrease during degradation.
[h] The absorption formed possibly by decreasing peak, 1782 cm⁻¹ and increasing peak, 1784 cm⁻¹.

5 FT-IR Studies of Composition of Epoxy Resins

Materials qualification requires methods of determining the composition and the degree of curing the resins. For glass-reinforced composites, the glass fraction is also required.

A variety of infrared methods have been available for the simple problems of verifying the purity of the initial resin components and mixtures of those components. The simplest possible test of the purity of a reactant is to compare its spectrum with a known standard. Subtractive methods may be employed to emphasize the most significant intensity changes [19]. Composition analysis of the blended resins and hardeners becomes more difficult. In some cases, one may simply select characteristic isolated absorption bands for each component and estimate the composition by comparison to standard spectra of known composition [20, 21].

Table 3. Least-Squares Analysis 2000–1400 cm^{-1} of 1:1 Stoichiometry Anhydride:Epoxide Mixtures (0.5% wt BDMA Catalyst) Crosslinked at 90 °C

Uncured	Least-squares calculation (%)	Composition by weighing	Extent crosslinked from I(1860)/I(1608)
EPON 628	51.99 = 1.89 (wt%)	50.77	
NMA	48.01 = 0.96 (wt%)	49.23	
Extent crosslinked	−1.65 = 1.37		0.0
30 min a 90 °C			
EPON 828	52.06 = 1.97	50.77	
NMA	47.94 = 1.00	49.23	
Extent crosslinked	5.97 = 1.42		6.6
60 min a 90 °C			
EPON 828	51.62 = 2.11	50.77	
NMA	48.38 = 1.07	49.23	
Extent crosslinked	12.42 = 1.51		13.2
90 min a 90 °C			
EPON 828	51.58 = 2.24	50.77	
NMA	48.42 = 1.14	49.23	
Extent crosslinked	18.47 = 1.60		18.8
120 min a 90 °C			
EPON 828	51.11 = 2.31	50.77	
NMA	48.89 = 1.17	49.23	
Extent crosslinked	28.14 = 1.64		31.0
6 hr a 90 °C			
EPON 828	51.99 = 1.89	50.77	
NMA	48.01 = 0.96	49.23	
Extent crosslinked	65.35 = 1.36		70.2

The spectra of the pure components can be used to determine the composition of the neat resin system. The sensitivity in FT-IR can be improved by using least-squares curve fitting of the digitized spectra of the neat resin [22]. A least squares analysis uses all of the spectral data in the region of interest and accounts for spectral overlapping features. It is also possible to use weighing factors to maximize use of the spectral data available and minimize those regions with high noise [23] and base lines can be calculated as well [24]. These techniques have been used for the determination of composition in epoxy resins [7,8]. A method for quality control of the epoxy resin system has also been described [25]. In Table 3, the results are shown for EPON 828/NMA for the uncured system and the agreement between calculated and experimental is excellent [25]. A quality control study was made of the spectra of the various components and the spectroscopic changes induced by storage and hydrolysis [25].

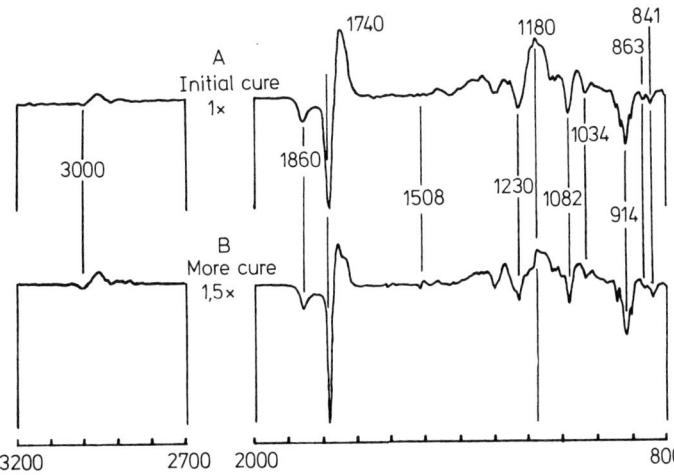

Fig. 1. Difference spectrum depicting the infrared spectral changes accompanying the 80 °C cross-linking of a stoichiometric mixture of EPON 828 and NMA with 2.0% weight BDMA catalyst

The determination of the degree of curing in epoxy resins for the cured samples has been examined. The use of the 918 cm^{-1} band [26] to measure the epoxy content and the hydroxy content [27] at 3,450 cm^{-1} is quite well documented. The spectral effects introduced by curing can be isolated by difference spectroscopy. The results are shown in Fig. 1 where the spectrum of an uncured epoxy resin is subtracted from the same material after curing at 66 °C for 22 hours. The criterion for subtraction is elimination of the 1,508 cm^{-1} band presumed to be associated with the benzene ring and therefore independent of the degree of cure but proportional to the mass of epoxy resin in the beam. The major features in this difference spectrum are as follows: intensity decreases at 1,860, 1,780, and 1,082 cm^{-1}, indicative of reaction of anhydride groups; intensity decreases at 1,034, 914, 863, and 841 cm^{-1}, indicative of a reduction in epoxy groups; and intensity increases at 1,740 and 1,180 cm^{-1} due to formation of ester linkages [25]. In Fig. 1(b), the spectrum represents a 96.2 percent cured sample (additional 160 min curing at 149 °C) minus the 58.8 percent cured

Table 4. Least-squares analysis of S-glass reinforced crosslinked epoxy matrix

Temp. time of curing	E (predicted)	Region			
		2000–1400 cm^{-1}		1550–850 cm^{-1}	
		R	E	R	E
80 °C, 70 min	31.1	1.2	51	0.93	54
80 °C, 90 min	39.3	1.1	59	0.88	55
80 °C, 110 min	47.5	1.1	57	0.90	59
80 °C, 130 min	55.6	1.0	62	0.92	62
80 °C, 180 min	63.7	1.0	70	0.96	67
160 °C, 30 min	83.5	1.0	82	0.93	80

sample. In this case, one observes that more anhydride has disappeared versus the amount of ester that has been formed. This difference is explained by the greater participation of the initiator in opening the anhydride ring. Thus the precise nature of the molecular structure depends on the extent and nature of the curing process.

When the spectra of the pure components are used to determine the composition of the cured resins, the results are shown in Table 4 [25]. In this case, the agreement between the expected results and the experimental results is unacceptable. Apparently, the curing process has introduced extensive spectral changes such that the spectra of the initial reactants do not represent their spectra in the cured state [25].

The analysis of any multicomponent resin or composite is greatly facilitated when the spectrum of that material is expressed by a linear combination of a finite set of "pure component" spectra. The entire process may be separated into three steps: calculation of the number of species present, identification of each of those species, and curve fitting of the spectra of these species to the spectra of the composites [7]. The technique for determining the number of components in the mixture is called factor analysis or major component analysis and has been described in detail in a number of publications [28,29]. Factor analysis is concerned with a matrix of data points. So, in matrix notation we can write the absorbance spectra of a number of mixtures as:

$$A = EC$$

where A is a normalized absorbance matrix which is rectangular in form having columns containing the absorbance at each infrared frequency recorded and the rows corresponding to the different mixtures being studied. The A matrix could thus be 400 × 10 corresponding to a measurement range of 400 cm^{-1} at one wavenumber resolution for 10 different mixtures. E is the molar absorption coefficient matrix and conforms with the A matrix for the frequency region but only has the number of rows corresponding to the number of absorbing components. C is the concentration matrix and has dimensions of the number of components by the number of mixtures being studied. We do not know E or C. Factor analysis allows one to determine both of them under certain circumstances [28].

Factor analysis allows initially a determination of the number of components required to reproduce the adsorbance or data matrix and the rank of A can be interpreted as being equal to the number of absorbing components. To find the rank of A, the matrix ATA is formed where AT is the transpose of A. This matrix termed the covariance matrix, has the same rank as A but has the advantage of being a square matrix with the dimensions corresponding to the number of mixtures being examined. In the absence of noise, the rank of A is given by the number of non-zero eigenvalues of M. Since the actual data contains noise and computational roundoff errors, additional nonzero eigenvalues (noise eigenvalues) will be generated by the computation. Theory shows that the eigenvalues can be grouped into two different sets: a set which contains the factors or components together with an error contribution and a secondary set composed entirely of error, if the log of the eigenvalues are plotted versus the number of the eigenvalues in descending order, a break will occur between the real and noise eigenvalues.

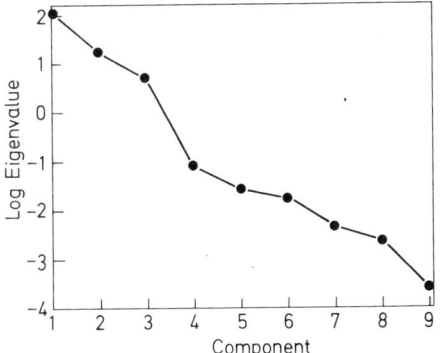

Fig. 2. Plot of log eigenvalue versus number of componenets showing number of components being 3

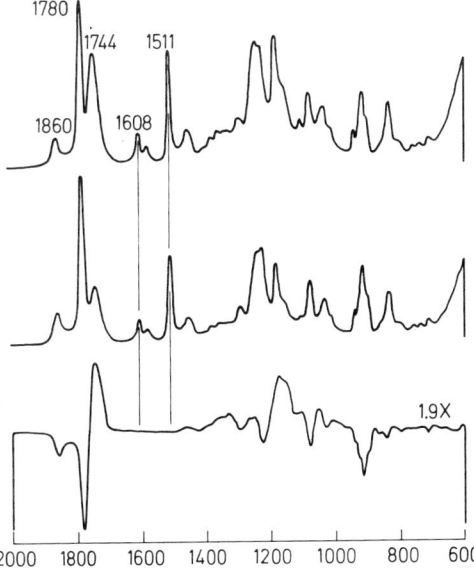

Fig. 3. Difference spectrum showing spectral changes occurring as a function of cure

Table 5. Least-Squares Analysis 2,000–1,400 cm^{-1} of 1:1 Stoichiometry Anhydride: Epoxide Mixtures (0.% wt BDMA Catalyst) [7]

Crosslinked at 95 °C			
Uncured	Least-squares calculation (%)	Composition bei weighing	Extent crosslinked from l(1860)/l(1608)
EPON 828	38.41 ± 1.06	34.04	
NMA	61.59 ± 0.54	65.96	
Extent crosslinked	−0.41 ± 0.59		0.0
60 min a 95 °C			
EPON 828	38.62 ± 1.34	34.04	
NMA	61.68 ± 0.68	65.96	
Extent crosslinked	6.97 ± 0.75		3.0
120 min a 95 °C			
EPON 828	37.11 ± 2.07	34.04	
NMA	62.89 ± 1.05	65.96	
Extent crosslinked	13.73 ± 1.14		9.4
6 hr a 95 °C			
EPON 828	37.08 ± 2.43	34.04	
NMA	62.92 ± 1.23	65.96	
Extent crosslinked	36.47 ± 1.34		38.7

Crosslinked at 90 °C			
Uncured	Least-squares calculation (%)	Composition by weighing	Extent crosslinked from l(1860)/l(1608)
EPON 828	51.99 ± 1.89 (wt%)	50.77	
NMA	48.01 ± 0.96 (wt%)	49.23	
Extent crosslinked	−1.65 ± 1.37		0.0
30 min a 90 °C			
EPON 828	52.06 ± 1.97	50.77	
NMA	47.94 ± 1.00	49.23	
Extent crosslinked	5.97 ± 1.42		6.6
60 min a 90 °C			
EPON 828	51.62 ± 2.11	50.77	
NMA	48.38 ± 1.07	49.23	
Extent crosslinked	12.42 ± 1.51		13.2
90 min a 90 °C			
EPON 828	51.58 ± 2.24	50.77	
NMA	48.42 ± 1.14	49.23	
Extent crosslinked	18.47 ± 1.60		18.8
120 min a 90 °C			
EPON 828	51.11 ± 2.31	50.77	
NMA	48.89 ± 1.17	49.23	
Extent crosslinked	28.14 ± 1.64		31.0
6 hr a 90 °C			
EPON 828	51.99 ± 1.89	50.77	
NMA	48.01 ± 0.96	49.23	
Extent crosslinked	65.35 ± 1.36		70.2

Factor analysis of the absorbance spectra of nine mixtures of NMA and EPON 828 (0.5 percent weight BDMA in each mixture also), each mixture having been crosslinked arbitrarily between 0 and 65 percent by heating at 80 °C was performed in the 2,000–1,400 cm^{-1} region. The plot of log eigenvalue in descending order shown in Fig. 2 shows a break in the eigenvalue magnitude between the third and fourth eigenvalue. Therefore, the spectrum of the crosslinked epoxy matrix may be approximated by a linear combination of only three linearly independent component spectra. The three components represent the spectra of the pure NMA, pure EPON 828, and a difference spectrum characteristic of the crosslinking reaction [7]. The difference spectrum was generated by subtracting a cured from an uncured spectrum as shown in Fig. 3. The results for the stoichiometric mixtures of epoxy and anhydride crosslinked to various extents are given in Table 5 for the 2,090–1,400 cm^{-1}. The agreement between the expected and observed data is excellent [7].

When factor analysis is applied to the fiber-reinforced composite, the results are indeterminate making the problem of estimating the number of components quite difficult. However, if one selects the spectra carefully, excellent results can be obtained including a determination of the fraction of glass [7, 8].

6 FT-IR Studies of Curing Kinetics and Mechanism of Epoxy Resins

In order to optimize the structure and properties of composites, a knowledge of the polymerization kinetics and mechanism are required. Various approaches have been taken to a determination of the kinetics of the polymerization including infrared spectroscopy [30, 31]. Although the various epoxide/anhydride/amine systems are

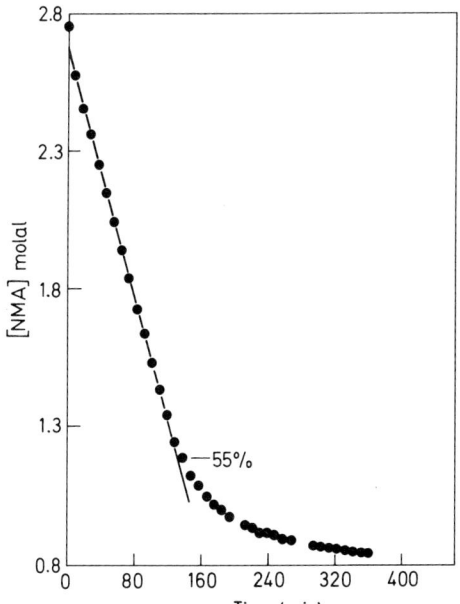

Fig. 4. Zeroth-order graph of a stoichiometric anhydride/epoxide tertiary amine-catalyzed copolymerization at 80 °C

extremely similar and in some cases identical, a generally accepted reaction mechanism is not available. Difficulties are encountered in the determination of the order of the reaction ranging from first to fourth order. Kinetic studies using FT-IR can be carried out in a thermally controlled cell installed in the IR instrument [9]. The reaction kinetics of NMA-cured EPON 828 was followed using the 1,858 cm^{-1} v_s(C=O) of the anhydride molecule. The polymerization was studied at 80 °C of a stoichiometric mixture of the reactants and the results are shown in Fig. 4 plotted as a zeroth-order process. Up to the gelation point (55%), the data is linear but slows above 55% and eventually stops at 71%. Plots of the nature can be easily generated from the FT-IR data for a variety of species since all of the frequencies are being recorded for every data point as opposed to the necessity in dispersion IR to follow only a single absorption mode.

7 FT-IR Studies of the Degradation Processes in Epoxy Resins

Understanding of the reversible and irreversible effects of moisture and other degradation processes is critical to the design of reliable composites [32]. FT-IR is particularly useful for this purpose as spectra can be obtained on the same or similar samples as a function of exposure time and very small spectral differences can be detected by absorbance subtraction. Moisture degradation is felt to occur by hydrolysis of the ester linkages. Such hydrolytic attack breaks the polymer chain creating two new end groups, a hydroxyl and a carboxyl (carboxylate under basic conditions). In concentrated HCl or NaOH solution at 50 °C, infrared spectra and specimen weight clearly indicate matrix hydrolysis [33].

FT-IR studies of adsorption of water in epoxy resins have been made [34]. The vibrations of the sorbed water within the epoxy film can be detected. The 1,628 cm^{-1} band is assigned to the bending mode of water; a broad v(OH) adsorption is also detected in the 3,700–3,000 cm^{-1}. Water in an epoxy resin exhibits a v(OH) mode which is as broad as that of liquid water but is more assymetric and stronger at higher frequencies. The bending mode of water in epoxy is observed at 1,628 cm^{-1} compared to the band in liquid water at 1,648 cm^{-1}. In general, the formation of hydrogen bonds affects the vibrational spectra of the groups involved by decreasing the frequency of stretching modes and increasing the frequency of bending modes. Interaction effects increase the infrared intensities as well. Since the bending mode of the sorbed water at 1,628 cm^{-1} is intermediate in frequency between the bands of liquid water and free water, it was proposed that the sorbed water is held within the resin by hydrogen bonding. For the most part, the polymer/water interactions are reversible for rather extended time periods [34].

The hydrolytic attack of water at the ester linkages in epoxies is accelerated in alkaline media and is a mechanically activated process [35]. Matrix hydrolysis is also enhanced in the presence of inorganic fillers [35].

8 Chemical Reactions of Epoxies with Coupling Agents

FT-IR has been used to study the nature of the interfacial reactions between epoxies and silane coupling agents in fiber-reinforced composites [36,37]. FT-IR has the capability of subtracting the absorbance of bulk resin and glass fiber allowing the spectra of the interfaces to be obtained, consequently, the difference spectrum allows the analysis of interface reactions. It is found that nadic methyl anhydride can react with gamma-methylaminopropyltriethoxysilane and N-methylaminopropyltrimethoxysilane. In comparing the relative reactivities of the two coupling agents to the epoxy resin, the secondary aminosilane has a higher reactivity than the primary aminosilane.

9 Basis of NMR Spectroscopy

Use of C-13 NMR in the solid state meant the solving of various spectroscopic problems. The solid state spectra were plagued with broad peaks due to strong dipolar interaction between the C-13 and protons and low sensitivity due to the low natural abundance of C-13 (1.1%). The C-13 T_1 relaxation times in the solid state are very long (minutes versus seconds for H-1) which made signal averaging a very time consuming task. High power decoupling, cross polarization and magic angle spinning were used to alleviate these problems.

9.1 Dipolar Decoupling

The line broadening due to C-13 and H-1 interaction and associated with long correlation times, can be removed with dipolar decoupling. Dipolar decoupling refers to strong proton decoupling [38,40]. This dipolar interaction can be eliminated with the irradiation of the proton at its Larmor frequency with a strong radio frequency field. The effect of this irradiation is that the H-1 nuclei become transparent to the C-13 nuclei by inducing rapid transitions between H-1 nuclei energy levels. In liquids, molecular tumbling motions average the dipolar interactions to zero. In solids, greater power levels are used because of the lack of these molecular tumbling motions.

9.2 Magic Angle Spinning

In solids, there is also chemical shift anisotropy which develops from nonspherical electron density around the C-13 nuclei. This nonspherical electron density is more predominant in aromatic and carbonyl types of structures. The carbon nuclei in these structures experience different shieldings from the magnetic field which is dependent on whether the bond axes are parallel or perpendicular to the applied field. All orientations between the parallel and perpendicular orientations are also possible taken with and without magic angle spinning. But, with magic angle spinning broadening by chemical shift anisotropy can be removed [41-45].

The magic angle of 54.7° with the field reduces $3\cos^2 - 1$, of residual static dipolar interactions and angular dependence of chemical shift anisotropy to zero [46]. The sample must spin at the magic angle in the order of the broadening caused by the chemical shift anisotropy. At high fields, C-13 and H-1 interaction are removed by decoupling.

9.3 Cross Polarization

Cross polarization can be used to increase the sensitivity of the dilute C-13 spins in solids. This technique transfers polarization from the abundant proton spin system to the dilute carbon spin system [47]. This is done by making a transformation to the rotating reference frame and matching the C-13 and H-1 energy levels by satisfying the Hartmann Hahn conditions, $w_H = w_C$ [47]. This process is much more rapid than the spin lattice relaxation process, therefore, the wait of 3 to $5T_1$ s of C-13 is not necessary between pulses when signal averaging. The other advantage is an intensity enhancement that is four times greater than an experiment without cross polarization because $\gamma_H/\gamma_C = 4$.

9.4 Solution Multinuclear NMR

Sojka and Moniz [48] used C-13 NMR solution spectroscopy to characterize diglycidyl ether of Bisphenol A (DGEBA) (Fig. 5). The hardeners used to cure the DGEBA were 1-piperidinyl-3-phenoxy-2-propanol and 1-ethoxy-3-phenoxy-2-propanol. The spectra in this study were obtained as a function of time and temperature. After combining the epoxy with a hardener, it was immediately placed in the probe. As the curing reaction proceeded, new signals were observed and monomer signals decreased in intensity. Model compounds were used for peak assignments. A generalized structure was proposed for the polymer.

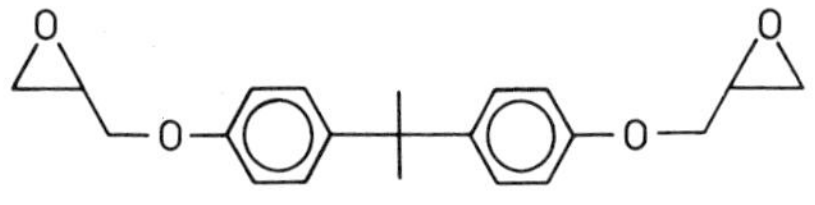

Fig. 5. Diglycidyl Ether of Bisphenol A DGEBA

Observation of resonances became difficult with the increase of molecular weight. To overcome this problem the probe temperature was increased to 180 °C which was above the T_g of the polymer. This procedure gave reasonable spectra from which unreacted epoxide groups were detected as well as extent of cure.

Kennedy, Guhaniyogi and Percec in 1982 [49] prepared glycidyl ethers of bisphenol-polyisobutylene and trisphenol-polyisobutylenes. These flexible epoxy prepolymers were examined. The epoxy prepolymers were divided into two segments. The first segment contained two or three epoxy groups. The second segment contained two or three "built-in" saturated hydrocarbon elastomer segments. It was noted that these polymers could have improved toughness, moisture, high temperature, and oxidative resistance. Proton NMR was used on model compounds to make peak assignments. End group analysis was done by UV spectroscopy and titration.

Gaul and Carr in 1983 [50] characterized an oxirane species which is reacted with Al(acac)$_3$ and diphenylmethyl silanol. C-13, Si-29 and Al-27 solution NMR was used along with model compounds to characterize the above reaction. The systems studied were 1,2-epoxybutane (1,2-EB), cyclohexene oxide and phenylglycidyl ether with Al(acac)$_3$ and diphenylmethyl silanol (DPMS).

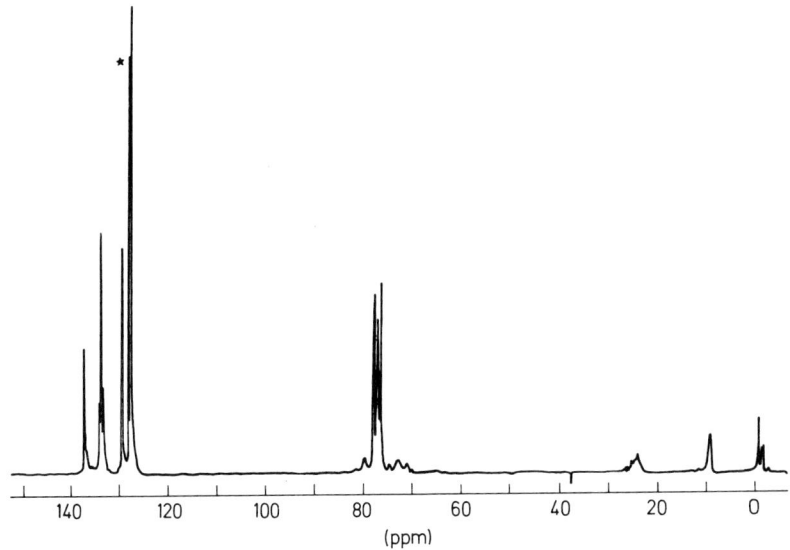

Fig. 6. C-13 NMR Spectrum of Reaction Mass of Heated Mixture of 1,2-epoxybutane, diphenyl-methyl silanol and Al(acac)$_3$ [50].

* Resonance signal in aromatic region (128.3 ppm) due to evolved benzene.

The spectra of the reacted 1,2-EB, DPMS and Al(acac)$_3$ is shown in Fig. 6. The C-13 chemical shifts are shown in Table 6. There is a signal at 128.3 ppm that does not appear in the unreacted silanol material. This peak is due to benzene [51]. This indicates phenyl cleavage of the silanol under the reactive conditions. The 1,2-EB and DPMS without Al(acac)$_3$ did not show any peak at 128.3 ppm which indicates that the aluminium species is responsible for phenyl cleavage and benzene formation.

A Si-29 spectrum of the same reaction is shown in Fig. 7. Figure 8 is the Si-29 spectrum of DPMS and Al(acac)$_3$ without 1,2-EB. The degree of complexity in the spectrum in Fig. 7 was said to be inconsistant with the formation of a copolymer. The similarity of the two spectra indicate that disproportionation of silanol material has occurred with formation of polysiloxanes. Studies by Hayes [52–54] along with the C-13 and Si-29 studies were said to point to the Al(acac)$_3$ chelate functioning as a disproportionation catalyst for the phenyl containing silanols.

Table 6. C-13 Chemical Shifts of 1,2-Epoxy Butane [50]

$$CH_3-CH_2-\overset{\overset{\displaystyle O}{\triangle}}{CH-CH_2}$$
 1   2   3   4

1	9.3 ppm
2	25.0 ppm
3	52.6 ppm
4	46.0 ppm

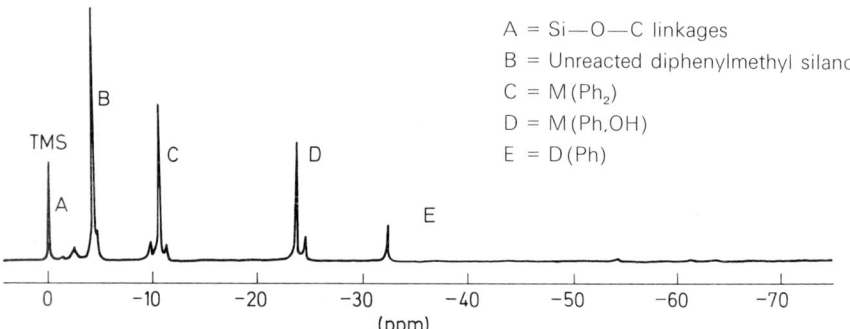

Fig. 7. Si-29 NMR spectrum of reaction mass of heated mixture of 1,2-epoxybutane, diphenylmethyl silanol and Al(acac)₃.
Note signals at "A" attributed to Si—O—C linkages [50]

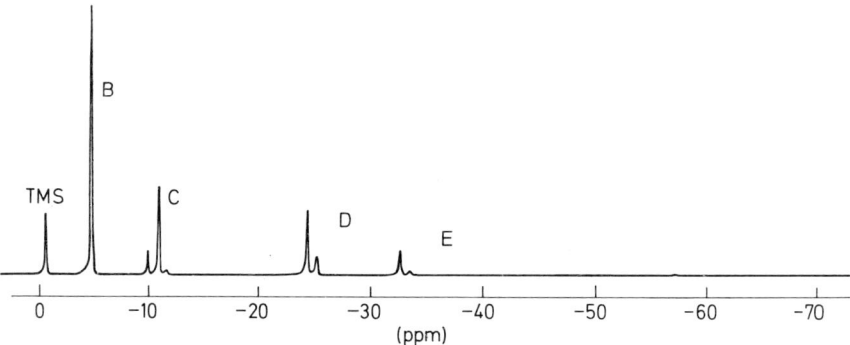

Fig. 8. Si-29 NMR spectrum of reaction mass of heated diphenylmethyl silanol and Al(acac)₃ heated in same manner as for reaction mass in Fig. 3.
Note lack of resonance signals at -1.0 to -3.0 [50]

The nature of the catalytically-active species was defined by studies done with Al-27 NMR. Al-27 NMR investigations are convenient because of advantageous nuclear properties [55]. The linewidths are dependent on molecular symmetry about the Al-27 nucleus because of its quadrupole moment. The highly symmetrical species such as Al(acac)₃ have the narrowest linewidths. If other ligands are introduced into the inner coordinating sphere of the aluminum atom, linewidths would increase in magnitude through the reduction of symmetry. The Al-27 NMR spectra of Al(acac)₃ had a single peak with a linewidth of 120 Hz. The spectrum of a mixture of DPMS and Al(acac)₃ showed little change in chemical shift but an increase in linewidth of 340 Hz was observed.

This research is moving toward the determination of experimental conditions necessary for the production of true silicone-epoxy copolymers. By using C-13, Si-29 and Al-27 NMR spectroscopy it was established that Al(acac)₃ is a disproportionation catalyst for phenyl containing silanols.

10 Resolution, Molecular Motion and Characterization of Epoxy in Solid State NMR

10.1 Molecular Motion in Solid State Proton NMR

The most studied epoxy resin in solid state NMR is diglycidyl ether of bisphenol A (DGEBA) shown in Fig. 5. Larsen and Strange, in 1973 [56], studied the molecular motion of the uncured DGEBA epoxy monomer and two higher molecular weight oligomers with pulsed proton NMR. Relaxation data were interpreted in terms of molecular motions occurring in the three epoxy resins. The proton spin-lattice (T_1) [57,58], rotating frame spin-lattice ($T_{1\varrho}$) [58], and spin-spin (T_2) [57] relaxation times were measured as a function of temperature. The temperature ranged from $-160°$ to 200 °C.

The T_1 relaxation time is dependent on molecular motion. T_1 can exhibit more than one minimum when measured as a function of temperature. This happens when several distinct motions occur simultaneously. The $T_{1\varrho}$ relaxation time is dependent upon molecular motion and has more than one minimum as well. The T_2 relaxation time is related to the inverse of the NMR linewidth.

Two T_1 minima were evident in the monomer. The low temperature minimum at -50 °C was attributed to methyl group reorientation. The high temperature minimum at 32 °C was said to be caused by molecular motion associated with the glass transition.

Two minima were observed in the $T_{1\varrho}$ relaxation time. The low temperature minimum at -117 °C was assigned to methyl group reorientation. The high temperature minimum at 8 °C was attributed to general molecular motion.

The T_2 relaxation time began to increase at approximately -5 °C. The increase in the T_2 relaxation time was also associated with molecular motion. T_2 approaches T_1 at higher temperatures.

Correlation frequencies determined from T_1, $T_{1\varrho}$ and T_2 relaxation times were plotted against reciprocal temperature and activation energies calculated. The methyl group had a comparatively high activation energy (4.7 kcal/mole) which was attributed to steric hindrance from the reorientation of the two methyls bound to the same carbon and steric hindrance arising from the two phenyl groups on the carbon atom.

The general molecular motion above the glass transition temperature had an activation energy of 33 kcal/mole. This data agrees with the 30 kcal/mole value obtained from NMR measurements by Clark-Monks et al. [59]. These measurements were from wide-line NMR. The T_2 and $T_{1\varrho}$ temperature dependent relaxation times gave an activation energy of 28 kcal/mole for the high temperature minimum. The slope of the log T_2 versus 1/T above 35 °C gave an activation energy of 11 kcal/mole. The monomer is a mobile liquid at this temperature. It was indicated that the T_1 was controlled by mainly translational motion of the molecules which is closely related to reorientational motion. The medium molecular weight oligomer resin was similar to the monomer but in the highest molecular weight oligomer resin the glass transition occurred at a much higher temperature. There was a third minimum for the $T_{1\varrho}$ which represented segmental motion. Table 7 gives a summary of the types of motion and their activation energies for the monomer and the two higher molecular weight mixtures.

Larsen and Strange [56] undertook the study of cured epoxy relaxation times and interpreted them in terms of molecular motion. These molecular motions were again divided into methyl group reorientation, segmental motion and general molecular motion. DGEBA was cured with 4,4'-methylenedianiline (MDA). The DGEBA and MDA were cured at three temperatures. The post cure was at 180 °C, the other two cures were at 100 °C and 54 °C. The T_1, $T_{1\varrho}$ and T_2 relaxation times were measured over a temperature range of -160 °C to the cure temperature of the sample.

Table 7. Activation energies for molecular motion in DGEBA [59]

Type of motion	Activation energies, kcal/mole	
	Monomer	Mixture II
Methyl group reorientation	4.7	—
Segmental motion	—	12–15
General molecular motion (Below 35 °C)		
From minima	33	27
From T_2 slope	28	28
General molecular motion (above 25 °C)		
From T_2 slope	11	—

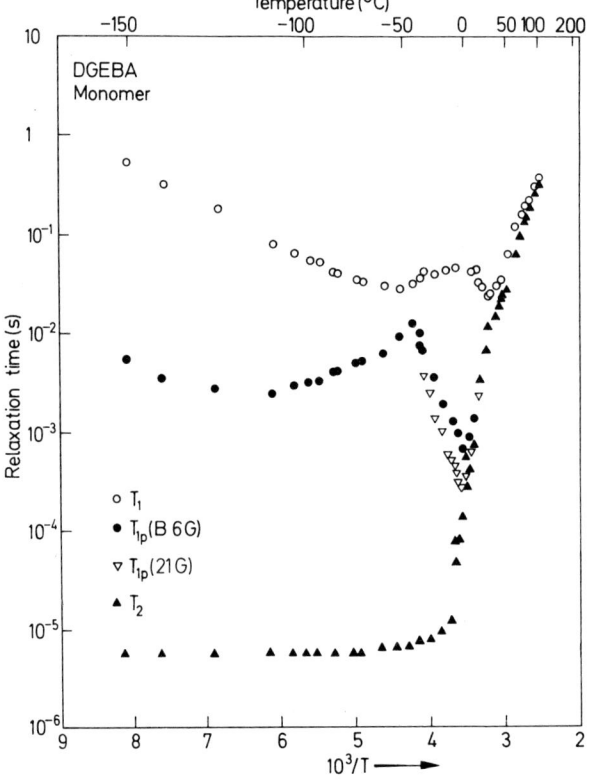

Fig. 9. Plots of log relaxation time versus reciprocal temperature for the monomer of DGEBA [56]

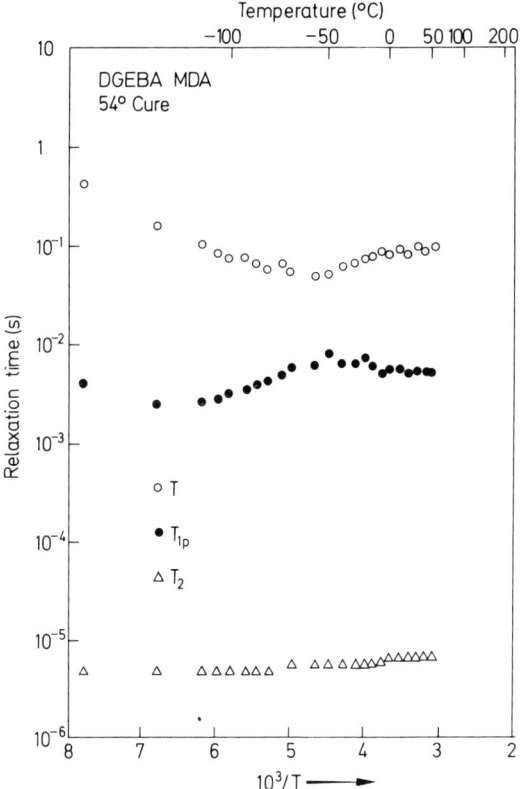

Fig. 10. Plot of log relaxation times versus reciprocal temperature for the diglycidyl ether of bisphenol. A cured with methylenedianiline at 54 °C [56].

In Fig. 9, the relaxation times are plotted versus reciprocal temperature for the 180 °C cure of DGEBA and MDA. The T_1 relaxation time minima at -70 °C as well as the $T_{1\varrho}$ relaxation time minima at -130 °C were attributed to methyl group reorientation.

General molecular motion was the cause of the T_2 increase at 185 °C. The minimums in the $T_{1\varrho}$ relaxation times at 26 °C and 90 °C were attributed to segmental motion. The T_1 relaxation time minimums were not clearly defined. The $T_{1\varrho}$ minimum for reorientation was not affected by the curing but the T_1 occurred at a higher temperature than in the cured sample. The implication is that the activation energy is higher for the cured resin than the uncured resin. The T_1 and $T_{1\varrho}$ minimums in the cured resin which represent segmental motion, showed no resolved T_1 or $T_{1\varrho}$ in the uncured resin. In the molecular motion associated with glass transition, only one difference was observed. The T_2 relaxation time had an abrupt increase in the cured resin. The 54 °C cure and the 100 °C cure relaxation plots are shown in Fig. 10 and 11.

The methyl group reorientation minima for the 54 °C cure was similar to the uncured resin whereas the 100 °C cure was fashioned after the 180 °C postcured resin.

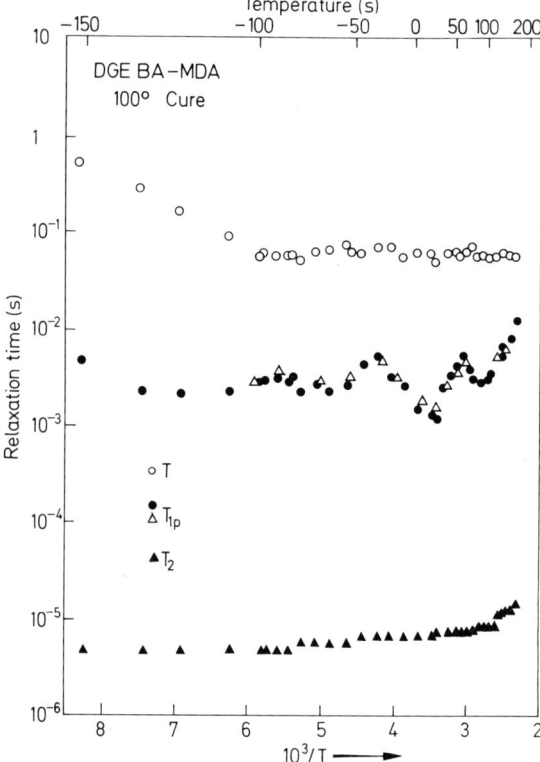

Fig. 11. Plot of log relaxation times versus reciprocal temperature for the diglycidyl ether of bisphenol. A cured with methylenedianiline at 100 °C [56]

In the 54 °C cure, the $T_{1\varrho}$ relaxation times were essentially independent of temperature where they should be dominated by segmental motion.

The relaxation data of the 100 °C cure had two well resolved minima which appear in the same place as the 180 °C postcure data. With curing, the slower segmental motion increases in intensity whereas the faster segmental motion decreases in intensity, which may be the result of crosslinking.

Mobility of the molecules at the 54 °C cure temperature was said to limit the extent of cure and prohibit annealing because of undefined segmental motion. At 100 °C curing temperature, the segmental motion becomes defined permitting annealing. The maximum crosslinking will appear at 180 °C curing temperature where the mobility of the system is at its maximum. At this temperature, the higher frequency segmental motions are limited by the tighter lattice. No spectra were exhibited in these studies.

10.2 Resolution and Mobility Studies in Solid State C-13 NMR

Resing and Moniz [60] cured DGEBA with three different curing agents, piperidine, hexahydrophthalic anhydride (HHPA) and nadic methyl anhydride (NMA). Spectra were collected with C-13 NMR using cross polarization and high power decoupling.

No magic angle spinning was incorporated making peak assignments difficult because of the peak broadening. But, when the cross polarization spectra were compared with a diagram, representing typical anisotropy patterns which were to be expected for the various carbon types, the peak assignments were made. These assignments represented protonated and unprotonated aromatic ring structures, a methylene carbon attached to an electronegative atom and aliphatic methyl and methylene groups.

The spectra of the cured epoxies were compared. It was found that the aliphatic region of the three spectra were distinctly different. The differences were associated with the rigidity of the polymer and the differences in the chemistry of the curing agents. Anisotropy in the aromatic region broadened in the order piperidine > HHPA > NMA. This was attributed to the motion in the backbone which reduces the anisotropy pattern. This should signify that the rigidities are in the order of piperidine > HHPA > NMA. The heat distortion temperatures did not give the same ordering.

It was felt that there was a large probability of chemical degradation, either near or at the glass transition temperature, taking place in the cured epoxy systems. This limitation was overcome by using solid state C-13 NMR [40].

It is evident that relaxation studies in the solid state can look at the motions which are responsible for the mechanical properties of the cured epoxy systems [43]. Therefore, Garroway, Moniz and Resing continued to do relaxation studies [61]. Garroway, et al. looked at four epoxy polymers based on the DGEBA resin. Two of the epoxy resins were cured with amines and the other two were cured with anhydrides. Proton enhanced spectra of the epoxy systems were generated. The solid state spectra were compared to the solution spectra of the unreacted epoxy. The epoxy resin of interest was again DGEBA which was reacted with:
1. Piperidine and carboxyl terminated butadiene acrylonitrile copolymer (CTBN) at 75 °C.
2. Meta phenylenediamine (MPDA) at 155 °C.
3. Hexahydrophthalic acid (HHPA) at 135 °C.
4. Nadicmethyl anhydride (NMA) with N,N-Dimethylbenzylamine.

It was thought that in the piperidine CTBN cured epoxy, cross polarization in the solid state spectra discriminated against the liquid-like lines of the CTBN. Only 4% piperidine was used in this sample. These spectra represented primarily the DGEBA polymer. When the spectra were compared to the solution spectra it was noted that the epoxy peak decreased in area with time of cure. The epoxide peak was not present in the solid state spectra. More rigidity was found in the MPDA cured polymer. This was based on heat distortion temperatures. This rigidity was thought to have broadened the peaks. The solid state spectra of the MPDA cured epoxy showed an absence of the epoxide peak indicating most of the epoxide groups were reacted.

The anhydride cured epoxies displayed a prominent carbonyl group of HHPA. The peak at 70 ppm was assigned to the products of the reacted epoxide groups. This peak was predominant in all four groups, piperidine-CTBN, MPDA, HHPA and NMA-DMBA.

In the NMA cured epoxy it appeared that the protonated aromatic lines were not broadened relative to the piperidine cure even though the heat distortion temperature of the NMA cured epoxy was greater.

It was noted that more refinement in spectral resolution was still needed. The combination of proton enhanced C-13 spectra combined with magic angle spinning made possible the identification of functional groups in the four DGEBA epoxy systems. A peak between 70–73 ppm were evidenced in all four systems indicative of the carboxyl-methine ether carbon of the reacted epoxide groups and adjacent methylene groups.

Garroway, Moniz and Resing [62] looked at high resolution in solid state C-13 NMR by using DGEBA cured with 5% piperidine as a model. It was decided rotating frame relaxation studies ($T_{1\varrho}$) would be measured and to estimate the spin-spin contribution to the relaxation rate. The $T_{1\varrho}$ of the carbon is defined at the observed rotating frame relaxation time constant. Figure 12 shows the C-13 NMR spectra of the PIP cured DGEBA. The dependence of the seven resolved peaks on spinning of the T_{CH} contact time is shown in Fig. 13. The rf field for this experiment was 38 kHz and the proton $T_{1\varrho}$ was equal to 2.6 ms. The nonprotonated carbon relaxation behavior showed averaging from magic angle spinning. Protonated carbons, because of strong interactions, relax in less time than one revolution of the sample specimen. havior showed averaging from magic angle spinning. Protonated carbons, because of strong interactions, relax in less time than one revolution of the sample specimen.

In Fig. 14, the rf dependence of the carbon $T_{1\varrho}$ times are shown. These $T_{1\varrho}$'s were normalized by T_{CH} values at 1 kHz from previous Fig. 13. Only data after 500 μs were used for determination of $T_{1\varrho}$. Only a very weak rf field dependence was seen. It was concluded that at room temperature and above 40 kHz fields that the C-13 $T_{1\varrho}$ values are determined by spin-lattice effects as well as by spin-spin events. The C-13 $T_{1\varrho}$ of oriented PE, at room temperature, even up to 80 kHz rf fields, are dominated by spin-spin effects.

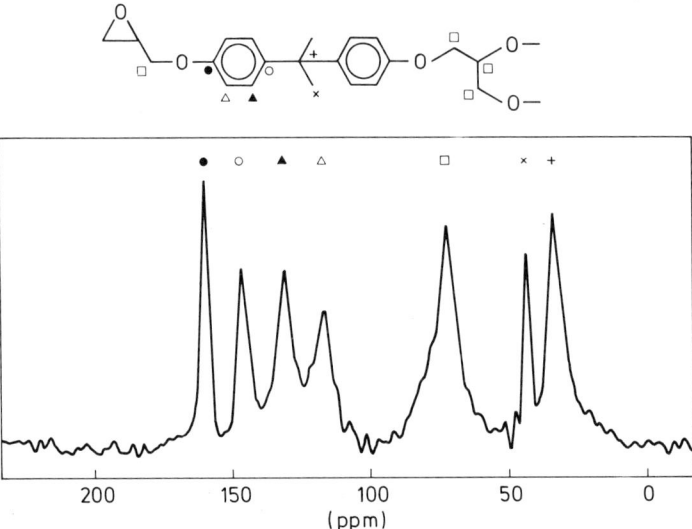

Fig. 12. C-13 spectrum of the piperidine (PIP) cured DGEBA epoxy polymer at room temperature. The assignments have been discussed and the structure indicates a possible polymerization mechanism [62]

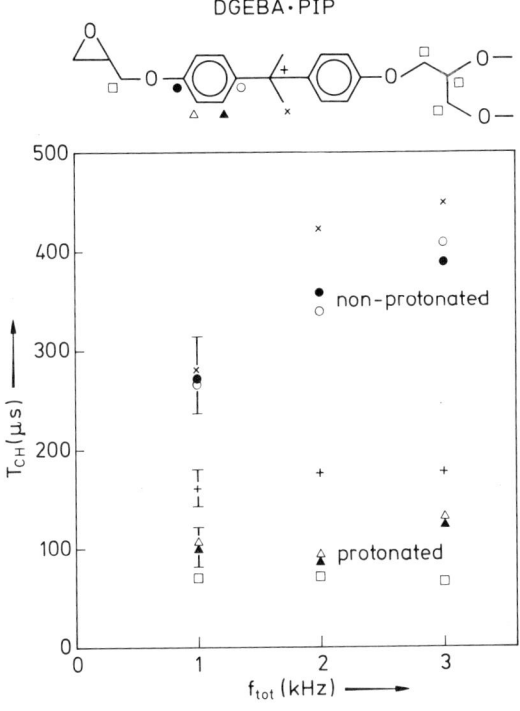

Fig. 13. Magic angle spinning alters the cross-polarization rates. Here T_{CH} is the time constant for spin lock cross-polarization under the Hartmann-Hahn condition at r.f. fields of 38 kHz [62]

At 55 kHz field, where relaxation times should indicate molecular motion, the relaxation times of the methyl groups showed a temperature dependence between -30 °C and 50 °C. An unresolved peak containing methylene and methine resonances showed a very weak temperature variation. Garroway et al. [62] concluded that the observed C-13 $T_{1\varrho}$ values for fields above 40 kHz were not dominated by spin-spin effects for the DGEBA-PIP system.

Moniz and Poranski [63] used both proton and C-13 NMR for the analysis of cured epoxy systems. These techniques were used in combination for the chemical characterization of cured systems and probing their molecular dynamics. The DGEBA epoxy system was a model for this characterization. The DGEBA resin used was a mixture of oligomers with various values of n. The epoxide equivalent weight (EEW) is a measure of the epoxide groups available for reaction and is an average for the mixture. Moniz described a technique using C-13 NMR to measure the EEW. Proton NMR methods have been used to measure the EEW in the past by Dorsey [64]. Hammerich and Willeboordse analyzed the precision of proton NMR EEW determinations [65]. The problems that Hammerich et al. found were inaccurate area measurements such as insufficient separation of main peaks on high EEW epoxy resins. The proton NMR results on low and medium EEW epoxy resins were satisfactory. The high EEW epoxy resins can be determined with more accuracy by the use of C-13 NMR which has larger chemical shift dispersion.

The basis for the EEW determination is the size of the oligomer structure, number of bridging carbons and the number of terminal epoxide groups. As n increases the number of bridging carbons increases but the number of epoxide groups stay the

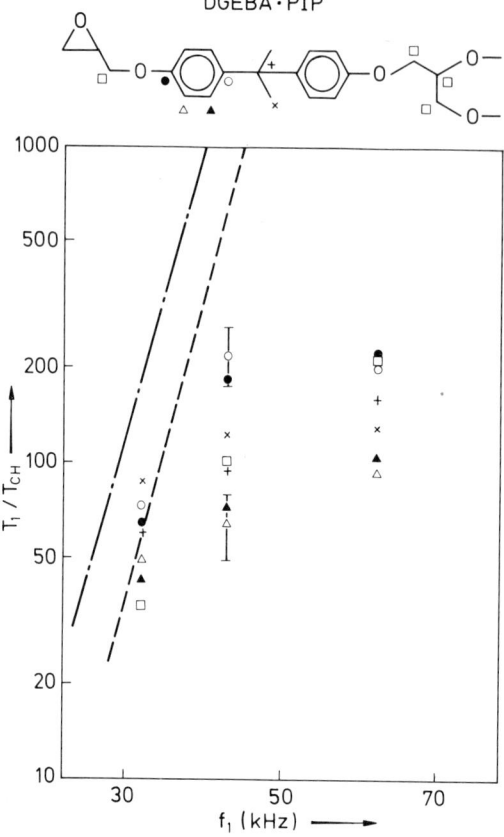

Fig. 14. R.f. field dependence of the C-13 T_1 times. The T_1 values have been normalized by T_{CH} (at 1 kHz spinning). The broken line estimates the field variation expected if the observed rotating frame relaxation were exclusively determined by spin-spin coupling. The dashed line represents the same field dependence and has been drawn through the 32 kHz data as an even more restrictive estimate; there is no evidence whether or not the low field data are determined exclusively by spin-spin effects. As the relaxation times at 43 and 66 kHz are shorter than those predicted for purely spin-spin effects, the high field results (and perhaps even at 32 kHz) indicate molecular motion [62].

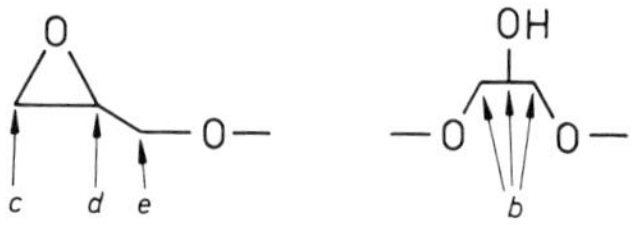

Fig. 15. Terminal ether carbons c, d and e. Bridge carbons b [63]

same. Ideally, the n-oligomer has a ratio of 2/3n of terminal ether carbons of types c, d and e to the bridge carbons b (Fig. 15). A set of hypothetical oligomer mixtures were generated and EEW was calculated by C-13 NMR and chemical analysis. Both methods gave equivalent results. The delay time used between pulses was determined by measuring the T_1s of the carbons in several epoxy systems. An adequate delay time was 35 s (5 times the longest T_1 of the carbon) [66]. Gated decoupling was used to suppress NOE. The results of the EEWs calculated from C-13 NMR and chemical analysis are show in Table 8.

Garroway et al. [67] used solid state C-13 NMR studies on epoxy polymers to

Table 8. EEW's found by carbon-13 and chemical analysis [63]

Resin	Carbon-13	Chemical	Literature [50]
DER 332LC	178	178	170–175
EPON 826	172	190	180–188
EPON 828[a]	176	194	185–192
EPON 828[b]	181	196	185–192
EPIKOTE	171	193	185–192
ARALDITE 7071	522	529	450–530
EPON 1001	568	600	450–530
EPON 1002	686	762	600–700
EPON 1004	971	1086	875–1025

[a] Manufactured in the United States;
[b] Manufactured in Canada

illustrate what properties can be extracted. These studies used cross polarization [47] with dipolar decoupling [38–40] and magic angle spinning [42–45]. The epoxy DGEBA, cured with piperidine, was used for the illustration of three stages of resolution (Fig. 16). In the top spectrum neither magic angle spinning nor high power decoupling were used. These conditions represent liquid state conditions. The middle spectrum added high power decoupling (proton rf field was 60 kHz). In the last spectrum magic angle spinning was implemented along with the high power decoupling. The chemical shift anisotropy is now an isotropic average and the signal-to-noise ratio is increased dramatically.

The DGEBA based epoxies were reacted with various curing agents. These spectra were than compared to the solution spectra of unreacted epoxy. It was shown that the peaks due to the epoxide carbons were not present in the polymerized DGEBA. In their previous study, Sojka and Moniz [48] showed that DGEBA partially cured with piperidine, exhibited a peak near the methylene peak that increased in intensity and complexity with the curing reaction. This peak was assigned to the carboxyl-methine ether carbon and the methylene carbon near the reaction site. These studies addressed the questions:
1. Are all carbons counted?
2. What are the limits of resolution?

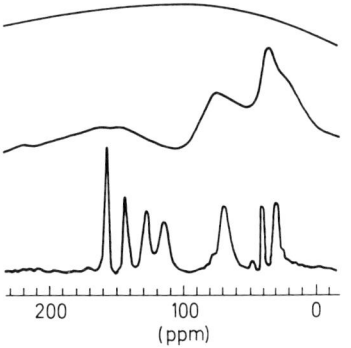

Fig. 16. Three stages of resolution in a C-13 spectrum of a cured epoxy. The top spectrum is obtained under conditions appropriate to a liquid-state spectrometer: no dipolar decoupling and no magic angle spinning. Dipolar decoupling at 60 kHz is used for the middle spectrum and to that is added magic angle rotation at 2.2 kHz for the bottom figure [67]

These questions were resolved with the use of the same relatively simple epoxy system. All C-13 nuclei in contact with the proton bath were counted when moderate spinning rates were used and in spin-lock cross polarization in rf fields not close to any $T_{1\varrho}$ minimum. The molecular motion determines the relaxation rate, under the Hartmann-Hahn condition when $T_{1\varrho} = T_2$. The spin-spin effects determine relaxation when $T_{1\varrho}$ does not equal T_2 under the same conditions [62]. The spin-spin fluctuations are in competition with the spin-lattice fluctuations in producing an effective relaxation time. To discriminate against the spin-spin fluctuations large rf fields are mandatory. It was pointed out that, with great care, C-13 NMR spectra can reflect molecular motion.

Balimann et al. [68] noted that one of the most obvious applications of solid state C-13 NMR is for insoluble or sparingly insoluble substances. Insoluble crosslinked epoxies fit this definition. Balimann looked at a diglycidyl ether based on bisphenol A which is called BADGE. BADGE was crosslinked with phthalic anhydride. The crosslinker is identified by aromatic and carboxylic carbons keeping the aliphatic region relatively simple making assignments of BADGE peaks fairly easy.

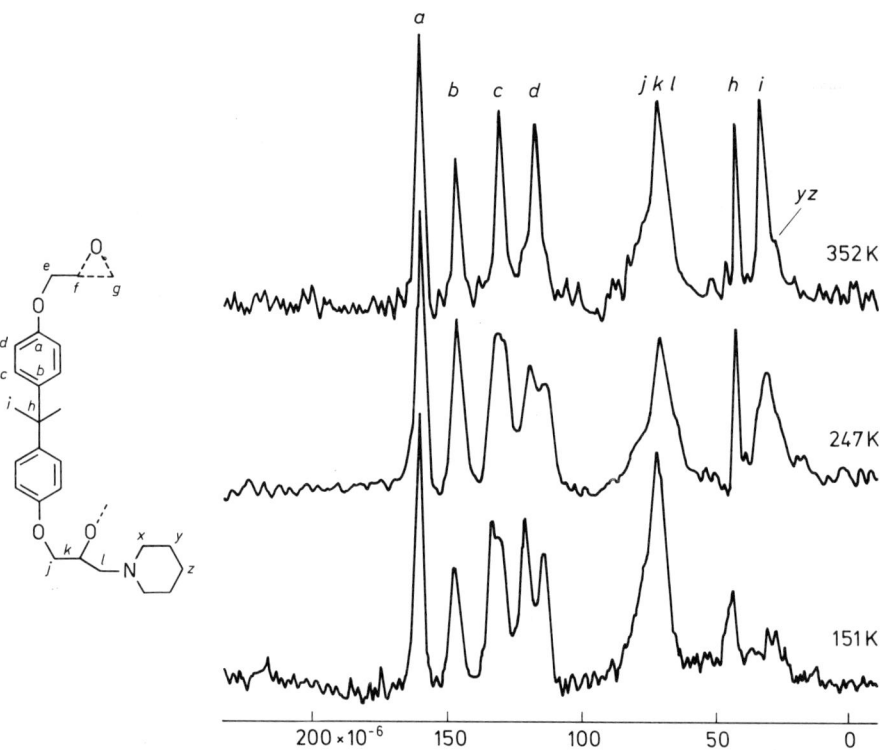

Fig. 17. The C-13 spectra over a 200 K temperature range for DGEBA epoxy cured with piperidine. The chemical structure on the left shows one half of the (symmetrical) monomer (top half) and a possible curing structure with piperidine (bottom half). In the spectra the methyl resonance broadens and then disappears at low temperature; the remaining peak at about 25×10^{-6} is assigned to the piperidine. The low-temperature splittings of peaks c and d collapse at higher temperature, indicating reorientation of the phenyl group with respect to the backbone [64]

VanderHart, Earl and Garroway [69] reported that the DGEBA $T_{1\varrho}$ values taken at 32 °C with an rf field of 66 kHz to be influenced predominantly by spin-lattice effects. From this information, the homogeneous contribution to the linewidth was attributed to the motional modulation of the C—H coupling. By comparing these values, it was noted that there was a 35% or less contribution to the linewidths of the cured epoxies attributed to motional modulation of the C—H coupling. The second homogeneous linewidth contribution came from the C-13 chemical shift anisotropy. It was noted that over half of the carbon linewidths exhibit chemical shift dispersion and are of inhomogeneous origin. This inhomogeneous line broadening was said to be contributed by inherent inequivalences of the carbons ortho to the oxygenated aromatic carbons. This inherent inequivalence was responsible for the broadening of the methyl resonances as well. Other inhomogeneous contributions arise from slight variations in bond angles, steric interactions and positions of anisotropic sources of molecular magnetic susceptibility.

At low temperatures the correlation times for motion are long, therefore, homogeneous broadening from relaxation does not play as important a part. It was pointed out that the line broadening at the low temperature was then due to inhomogeneous broadening. The high temperature spectra showed better resolution because the molecular motion averaged the chemical shift dispersion. An exception to the above data was the methyl resonance. By lowering the temperature, the methyl group goes from very fast motion to motional frequencies on the order of w_{1H}. The lower temperature broadens the line width of the methyl group because of the change in motion. Figure 17 points out that both the observed linewidth and the corresponding $(T_{1\varrho})^{-1}$ contributions have similar temperature dependence.

Garroway, et al. [70] defined problems due to dipolar spin-spin coupling which degrades resolution as well as interfering with the information about molecular motion. Spectra was compared (Fig. 18) of DGEBA polymerized with piperidine, an amorphous resin, a polycrystalline resin and the starting resin in the liquid state. A comparison of the liquid and crystalline phase resins showed the crystalline phase to be more complex. The magnetically inequivalent rigid carbons of the main peaks become equivalent in the solution state. The amorphous epoxy spectra has structural information but it is not as distinct as the crystalline.

In the polymerized epoxy spectra, the opening of the epoxide groups shifted the two epoxide resonances downfield. The aromatic peaks in the spectra of all phases remained identical. The crystalline environment has extreme regularity therefore lines are narrow. The methyl resonances were broadened. This was attributed to magnet inhomogeneity and magic angle imperfections. The amorphous stage had a distribution of local environments which quenched the structural detail. The polymerized epoxy froze in the random distribution of orientations decreasing resolution further.

Further evidence for isotropic chemical shifts was found in variable temperature studies. The aromatic carbon ortho to the oxygen resolved into two peaks at low temperatures. The aromatic carbons meta to the oxygen have two peaks, but these peaks are not as well resolved. As the temperature was raised, the splittings in both coalesced into one peak, indicating that there is rapid sampling from two magnetic environments. This is indicative of the motion of the phenyl group with respect to the backbone. At lower temperatures, the methyl resonance broadens and eventually

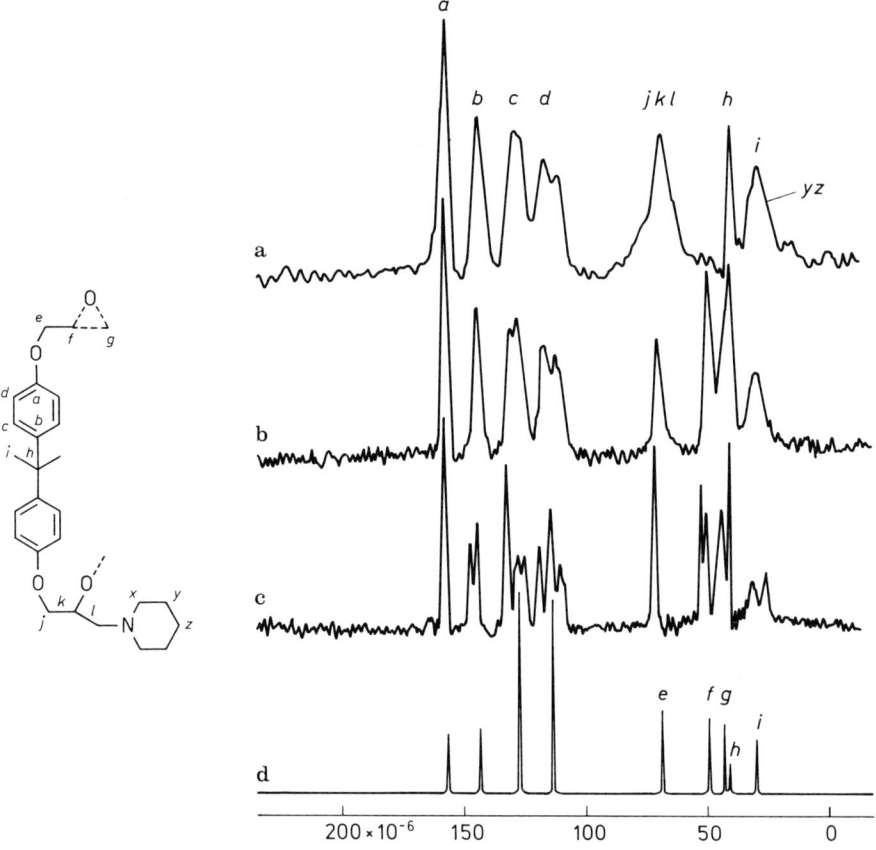

Fig. 18a–d. C-13 spectra of the epoxy resin DGEBA in four different phases:
(a) polymerized with 5% by mass of piperidine, 247 K.
(b) amorphous resin, 230 K.
(c) polycrystalline resin, 230 K.
(d) in CCl$_4$ solvent peak deleted.
The chemical structure on the left shows one half of the (symmetrical) monomer in the top half and a possible curing structure with piperidine (bottom half) in the aromatic region. In the amorphous phase much of this detail is smoothed out, presumably by the distribution of molecular environments leading to a distribution of isotropic chemical shifts. The spectra of amorphous and polymerized phases are quite similar in the aromatic region [69]

disappears. This was attributed to the increasing inefficiency of the proton decoupling as the methyl reorientation slows down from its ambient temperature rate. The chemical shift anisotropy was modulated by the phenyl and methyl motions reducing the efficiency of magic angle spinning.

Garroway, Ritchie, and Moniz [71] continued the characterization of epoxy polymers with respect to molecular motion using variable temperature, solid state C-13 NMR. DGEBA was the epoxy of interest. The DGEBA was cured with piperidine, m-phenylene diamine, hexahydrophthalic anhydride and nadic methyl anhydride. The piperidine cured DGEBA had the best resolved polymer spectra. This system was examined in detail. Peak assignments were from solution NMR [48, 72]. The cured epoxy polymer spectra were compared to the spectra of the epoxy monomer. These monomers had two forms, one form was amorphous and the other was crystalline. The crystalline epoxy monomers were used in x-ray studies [73] as well. This study found evidence of two conformations available to the epoxy groups on one end of the molecule. The other end of the molecule has only one conformation.

The Grant-Cheney steric hindrance model [74] to estimate the C-13 splittings were applied. The evidence was said to be ambiguous but they felt that there definitely are two forms of the monomer present.

The carbons ortho to the oxygen split into two peaks at low temperatures. The carbons meta to the oxygen make a weak attempt to split. At high temperatures, the resonances of these two peaks merged into one line which continued to narrow

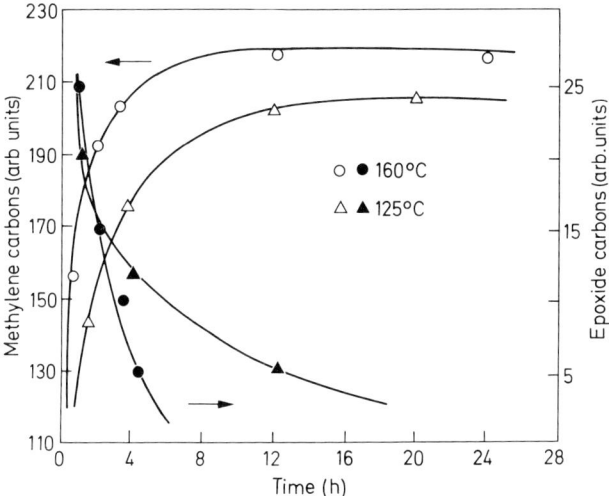

Fig. 19. Plot of peak intensities of methylene, epoxide carbon intensities against time [75]

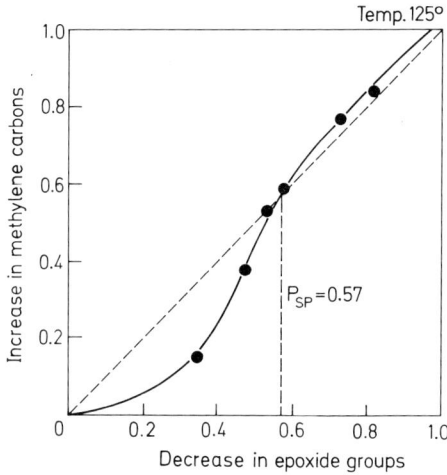

Fig. 20. Plot of increase in methylene resonance against the decrease in the epoxide group [75]

as the temperature increased. These resonances were said to imply the chemical exchange process in liquids. The spectra resolution of the glassy cured polymers was found to increase with temperature [69, 70].

Crystals of DGEBA monomer were prepared. The polycrystalline powder was run at different temperatures. There was no coalescence of the aromatic peaks.

Cholli, Ritchey, and Koenig [75] characterized the epoxy system DGEBA with solid state C-13 NMR. This study followed the disappearance of intensity of the epoxy carbons and the increase in intensity of the methylene carbons attached to the oxygen atoms. In Fig. 19, these intensity changes are plotted versus the curing time of the epoxy resins. Figure 20 plots the increase in methylene carbon intensities versus the decrease in epoxide group intensities.

The conversion of one epoxide group should give one oxymethylene unit leading to a straight line with a slope of one. This was not the case. There was deviation from ideality below 0.57 extent of reaction.

The behaviour in Fig. 20 was analyzed on the basis of chemical reaction during cure and the physical state of the system as a function of curing. During the early stages of curing, products may form from the epoxide groups reacting with H_2O, alcohol or HX to form products other than the oxymethylene units. Side reactions may also be caused by impurities. The by-products are felt to be less than 3% and should not be responsible for the deviation from ideal behavior.

This deviation from ideality may be caused by a change in the physical state of the system. This change in the state being induced during curing.

As crosslinking increases with curing, the system becomes more rigid, restricting the molecular motion. This restriction in motion affects the efficiency of the cross-polarization between the protons and carbons. This in turn affects the cross-relaxation time constants.

The ideal behavior after 0.57 extent of reaction may indicate the gel point. Dusek's theory [76-78] of cascade processes provides the changes in structural parameters with conversion. This along with C-13 NMR data in Fig. 20 shows the functionality of the epoxide as 2.6.

11 References

1. Griffiths, P. R.: Chemical Infrared Transform Spectroscopy, New York, Wiley, 1975
2. Ferraro, J. R., et al.: Fourier Transform Infared Specetroscopy Applications to Chemical Systems, Academic Press, New York, 1978, J. R. Ferraro, L. J. Basile, Eds.
3. Koenig, J. L.: Acct. of Chem. Res., *14*, 171 (1981)
4. Bell, R. J.: Introductory Fourier Transform Spectroscopy, Academic Press, New York, 1982
5. Griffiths, P. R.: "FT-IR Theory and Instrumentation", in: Transform Techniques in Chemistry, P. R. Griffiths, Ed., Plenum Press, 1978
6. Cooley, J. W., Tukey, J. W.: Math. Comput. *19*, 297 (1965)
7. Antoon, M. K., Zehner, B. E., Koenig, J. L.: Polym. Comp. *1*, 24 (1980)
8. Antoon, M. K., Zehner, B. E., Koenig, J. L.: Polym. Comp. *2*, 81, (1980)
9. Antoon, M. K., Koenig, J. L.: J. Polym. Sci., Polym. Chem. Ed., *19*, 549 (1981)
10. Koenig, J. L.: Adv. Polym. Sci., *54*, 87 (1983)
11. Dannenberg, H.: SPE Trans., 78–88, Jan., 1963
12. Perkinson, J. L.: A Method of Obtaining a True Infrared Spectrum of the Epoxy from a Cured Fiber Glass-Epoxy Composite, Gov., Rep. Announce (U.S.), *71*, No. 19, 88, 1971
13. Graf, R., Koenig, J. L., Ishida, H.: Appl. Spectrosc., (in press)

14. Fuller, M. P., Griffiths, P. R.: Anal. Chem., *50*, 1906 (1978)
15. Kirshnan, K.: Appl. Spectrosc., *35*, 549 (1981)
16. Lee, H., Neville, K.: Handbook of Epoxy Resins, New York, McGraw-Hill, 1967
17. Antoon, M. K.: Ph. D. Dissertation, Case Western Reserve University, Cleveland, Ohio, Appendix I, 1981
18. Pearce, E. M., Bulkin, B. J., Lin, S. C.: J. Polym. Sci., Polym. Chem. Ed., *17*, 3121 (1979)
19. Murphy, W. J.: Ed., Ind. Eng. Chem., *15*, 609 (1943)
20. Barnes, R. B.: Applications of Instruments in Chemistry, eds. R. E. Burk and O. Grummitt, Interscience, New York, p. 144, 1945
21. Murphy, W. J. Ed.: Ind. Eng. Chem., *15*, 659 (1943)
22. Antoon, M. K., Koenig, J. H., Koenig, J. L.: Appl. Spectrosc., *31*, 518 (1977)
23. Haaland, D. M., Easterling, R. G.: Appl. Spectrosc., *34*, 539 (1980)
24. Haaland, D. M., Easterling, R. G.: Appl. Spectrosc., *36*, 665 (1982)
25. Antoon, M. K., Starkey, K. M., Koenig, J. L.: Applications of Fourier Transform Infrared Spectroscopy to Quality Control of the Epoxy Matrix, Composite Materials, Fifth Conference, ASTM, p. 541, 1979
26. Dannenberg, H., Harp, W. R. Jr.: Anal. Chem., *28*, 86 (1956)
27. Patterson, W. A.: Anal. Chem., *26*, 823 (1954)
28. Malinowski, E. R., Lowery, D. G.: Factor Analysis in Chemistry, John Wiley, New York, 1980
29. Antoon, M. K., D'Esposito, L., Koenig, J. L.: Appl. Spectrosc., *33*, 351 (1979)
30. Luston, J., Manasek, A., Kulickova, M.: J. Macromol. Sci., Chem. *12*, 995 (1978)
31. Feltzin, J., Longenecker, D. M., Petker, I.: S.P.E. Trans., 111–116, April, 1965
32. Antoon, M. K., Koenig, J. L.: J. Macromol. Sci., Rev. Macromol Chem., *C19*, 135 (1980)
33. Hojo, H., Tsuda, K.: Effects of Chemical Environments and Stress on Corrosion Behaviors of Glass Fiber Reinforced Plastics and Vinyl Ester Resin, Proc., 34th Ann. Tech. Conf., Reinf. Plast./Compos. Inst., SPI, Sec. 13-B, 1979
34. Antoon, M. K., Koenig, J. L., Serafini, T.: J. Polym. Sci., Polym. Phys. Ed., *19*, 1567 (1981)
35. Antoon, M. K., Koenig, J. L.: J. Polym. Sci., Polym. Phys. Ed., *19*, 197 (1981)
36. Chiang, C. H., Koenig, J. L.: Polym. Comp., *1*, 88 (1980)
37. Chiang, C. H., Koenig, J. L.: Polym. Comp., *2*, 192 (1981)
38. Bloch, F.: Phys. Rev., *111*, 841 (1958)
39. Sarles, L. R., Cotts, R. M.: Phys. Rev., *111*, 853 (1958)
40. Pines, A., Gibby, M. G., Waugh, J. S.: J. Chem. Phys., *59*, 569 (1973)
41. Andrew, E. R.: Phil. Trans. Roy. Soc. Lond., *A299*, 29 (1981)
42. Andrew, E. R.: Arch. Sci., (Geneva), *12*, 103 (1959)
43. Andrew, E. R.: Prog. NMR Spectros., *8*, 1 (1971)
44. Lowe, I. J.: Phys. Rev. Lett., *2*, 285 (1959)
45. Kessemeier, H., Norberg, R. E.: Phys. Rev., *155*, 321 (1967)
46. Mag. Res. Rev., *7*, No. 2, 87 (1982)
47. Hartmann, S. R., Hahn, E. L.: Phys. Rev., *128*, 3042 (1952)
48. Sojka, S. A., Moniz, W. B.: J. Appl. Polym. Sci., *20*, 1977 (1976)
49. Kennedy, J. P., Guhaniyogi, S. C., Percec, V.: Polym. Bull., *8*, 571 (1982)
50. Gaul, J. H. Jr., Carr, T. M.: Spectros. Lett., *16*, 9, 651 (1983)
51. Strothers, J. B.: Carbon-13 NMR Spectroscopy, New York: Academic Press, p. 311, 1972
52. Hayes, S., et al.: J. Polym. Sci., Polym. Chem. Ed., *19*, 9, 2185 (1981)
53. Hayes, S., et al.: J. Polym. Sci., Polym. Chem. Ed., *19*, 11, 2977 (1981)
54. Williams, E. A., Cargioli, J. D.: "Silicon-29 NMR Spectroscopy" Annual Reports in NMR Spectroscopy, *9*, Academic Press, Lond., G. A. Webb, Ed., p. 222, 1979
55. Hinton, J. A., Briggs, R. W.: NMR and the Periodic Table, Academic Press, R. K. Harris, and B. E. Mann, Eds., Chap. 9, p. 279, 1978
56. Larsen, D. W., Strange, J. H.: J. Polym. Sci., Polym. Phy. Ed., *11*, 65 (1973)
57. Slichter, W. P.: NMR, Basic Principles and Progress, *4*, Springer Berlin, p. 209, 1971
58. Connor, T. M.: NMR, Basic Principles and Progress, *4*, Springer Berlin, p. 244, 1971
59. Clark-Monks, C., Ellis, B.: J. Polym. Sci., *8*, A-2, 2203 (1970)
60. Resing, H. A., Moniz, W. B.: Macrom., *8*, 4, 560 (1975)
61. Garroway, A. N., Moniz, W. B., Resing, H. A.: Am. Chem. Soc., Coating and Plastics Preprint of the 172nd Meeting, *36*, No. 2, 1976

62. Garroway, A. N., Moniz, W. B., Resing, H. A.: Pulsed Nuclear Magnetic Resonance in Solids, The Faraday Disc. Chem. Soc., Lond., No. 13, p. 63, 1978
63. Moniz, W. B., Poranski, C. F., Jr.: Epoxy Resin Chemistry, Chap. 7, ACS Symp. Ser., No. 114, Washington D.C., Amer. Chem. Soc. 1979
64. Dorsey, J. G., Dorsey, G. F., Rutenberg, A. C., Green, L. A.: Anal. Chem., 49, 1144 (1977)
65. Hammerich, A. D., Willeboordse, F. G.: Anal. Chem., 45, 1696 (1973)
66. Shoolery, J. N.: Prog. NMR Spectros., 11, 79 (1977)
67. Garroway, A. N., Moniz, W. B., Resing, H. A.: Carbon-13 NMR in Polymer Science, ACS Symp. Ser., No. 103, W. M. Pasika, Ed., 1979
68. Balimann, G. E., Groombridge, C. J., Harris, R. K., Packer, K. J., Say, B. J., Tanner, S. F.: Phil. Trans. R. Soc. Lond., A299, 643 (1981)
69. VanderHart, D. L., Earl, W. L., Garroway, A. N.: J. Magn. Reson. 44, 361 (1981)
70. Garroway, A. N., VanderHart, D. L., Earl, W. L.: Phil. Trans. R. Lond., A299, 609 (1981)
71. Garroway, A. N., Ritchey, W. M., Moniz, W. B.: Macromolecules, 15, 1051 (1982)
72. Poranski, C. F., Jr., Moniz, W. B., Birkle, D. L., Kopfle, J. T., Sojka, S. A.: NRL Report No. 8092, Naval Research Laboratory, Wash. D.C., 1977
73. Flippen-Anderson, J. L., Gilardi, R: Acta Crystallogr., B37, 1433 (1981)
74. Grant, D. M., Chaney, B. V.: J. Am. Chem. Soc., 89, 5315 (1967)
75. Cholli, A., Ritchey, W. M., Koenig, J. L.: Preprint Am. Chem. Soc., 1984
76. Dušek, K., Ilavský, M., Luňák, S.: J. Polym. Sci., Polym. Symp. 53, 29 (1975)
77. Luňák, S., Dušek, K.: J. Polym. Sci., Polym. Symp. 53, 45 (1975)
78. Dušek, K.: in Rubber-Modified Thermoset Resins, ACS Adv. Chem. Ser. 208, eds. Riew, C. K., Gillham, J. K., Washington: Amer. Chem. Soc. (1984) p. 3

Editor: K. Dušek
Received Mai 20, 1985

Kinetics, Thermodynamics and Mechanism of Reactions of Epoxy Oligomers with Amines

B. A. Rozenberg
Institute of Chemical Physics, Academy of Sciences of the U.S.S.R.
142432 Chernogolovka, Moscow Region/U.S.S.R.

In this review, modern interpretations of the kinetics, thermodynamics and mechanisms of curing of epoxy oligomers with primary, secondary, and tertiary amines and their mixtures, as well as the structure of the resulting polymers are discussed. The effect of the structure of the reagents on their reactivity is analyzed. Kinetic peculiarities of the deep stages of the curing process are emphasized Problems to be solved in the future are formulated.

1 Introduction		115
2 Primary and Secondary Amines		115
2.1	Catalytic Reaction	115
2.2	Noncatalytic Reaction	116
2.3	Addition Mechanism	118
2.4	Reaction Mechanism	119
2.5	Reaction Thermodynamics	120
	2.5.1 Donor-Acceptor Complexes	120
	2.5.2 Change of the Reaction Enthalpy	125
2.6	Reaction Kinetics	127
	2.6.1 General Kinetic Principles	127
	2.6.2 Structure of the Effective Reaction Rate Constant	128
	2.6.3 Compensation Mechanism	129
	2.6.4 Substitution Effect	130
2.7	Kinetic Features of the Deep Conversion Stages	133
	2.7.1 Autoinhibition Effect	133
	2.7.2 Diffusion Control of the Curing Reaction	135
	2.7.3 Bimolecular Reaction Rate Constant for the End Functional Groups During Network Formation	137
	2.7.4 Topological Reaction Limit	137
	2.7.5 Reaction Kinetics and Polymer Structure	138
2.8	Curing of Epoxy-amine Systems at Nonisothermal Conditions	139
2.9	Reagent Structure and Reactivity	140
	2.9.1 Amines	140
	2.9.2 Epoxy Compounds	141
	2.9.3 Reagent Structure and Ineffective Cyclization Reactions	142
2.10	Effect of Solvent Nature	144
2.11	Side Reactions	144

3 Tertiary Amines 146
   3.1 Main Kinetic Principles 147
   3.2 Polymer Structure 148
   3.3 Reaction Mechanism 149
        3.3.1 Initiation 150
        3.3.2 Chain Propagation 154
        3.3.3 Chain Termination Reactions 156

4 Amine Mixtures 158

5 Conclusion 160

6 References 161

1 Introduction

Technical applications of epoxy resins started about fifty years ago and have resulted in a number of remarkable achievements in the filed of high-strength polymer composite materials, lacquers, coatings, adhesives, etc. [1-11].

Amines were one of the first hardeners of epoxy resins [1, 12], and at present they retain their leading position among all known hardeners of this type. The amine hardeners will most likely also be used in the future because they are fairly accessible, highly reactive, and their properties can be readily modified. The mechanical properties of the cured resins obtained are far better than those of the known polymer binders [2-4], they have high dielectric characteristics, chemical resistance, etc.

During the last twenty years, curing of epoxy resin with amines and other hardeners has been extensively investigated. New fundamental results were obtained in the chemistry of epoxy compounds along with a better understanding of the molecular, topological structure and properties of the cured epoxy resins [2-6, 13-15]. These were useful in developing the kinetic models of the curing processes and helpful to investigate the scientifically substantiated principles for the selection of the time-temperature curing regime [5, 16-18]. Besides, a knowledge of the curing mechanism provides a basis for the rational selection of a curing system and for the control of the structure of the three-dimensional network.

A great number of investigations has been devoted to the chemistry of epoxy compounds and clarification of the mechanism of curing of epoxy oligomers with various hardeners, summarized in monographs [1, 12, 19-23] and reviews [24-25], which have become, however, somewhat obsolete.

Low-molecular-weight model compounds such as phenylglycidyl or other monoglycidyl ethers as well as primary, secondary and tertiary amines have been used for the study of the kinetics, thermodynamics and mechanism of curing. To reveal the kinetic features of network formation, results of studies of the real epoxy-amine systems have also been considered. Another problem under discussion is the effect of the kinetic peculiarities of formation of the epoxy-amine polymers on their structure and properties.

Since the mechanism of curing of epoxy compounds with primary and secondary amines differs essentially from that using tertiary amines, we shall consider this case separately.

2 Primary and Secondary Amines

2.1 Catalytic Reaction

In curing of epoxy oligomers with primary and secondary amines, the following main reactions take place[1]:

$$RNH_2 + PhOCH_2CH\underset{O}{-}CH_2 \rightarrow RNHCH_2\underset{OH}{C}HCH_2OPh \qquad (1)$$

$$(A_1) \qquad (E) \qquad (A_2)$$

---

[1] To simplify the presentation, the reactions of the monofunctional compounds are shown

$$A_2 + E \rightarrow RN \underset{CH_2\underset{|}{C}HCH_2OPh}{\overset{CH_2\underset{|}{C}HCH_2OPh}{\diagup}} \quad \begin{array}{c} OH \\ \\ OH \end{array}$$

$$(A_3)$$

where R is aryl or alkyl.

A tertiary amino group formed in curing with aliphatic amines can sometimes catalyze the epoxy group polymerization. When aromatic amines are used as curing agents. such reactions do not take place at all.

First we shall consider only the mechanism of reactions proceeding within the normally used curing temperature range (up to about 150 °C). At higher temperatures, a number of side reactions can take place, which will be considered below.

It is well known [1, 13, 14, 20–25] that the addition of hydroxyl-containing compounds (water, alcohols, phenols, acids) considerably promotes the interaction of epoxy compounds with amines and other nucleophilic reagents. In this case, the epoxy ring carbon atom becomes more sensitive to nucleophilic attack. The reaction proceeds through a trimolecular transition state initially suggested by Smith [26, 27] for the reactions of epoxy compounds with amines²

$$\begin{array}{c} \diagdown | \diagup \\ N \\ \vdots \\ PhOCH_2CH-CH_2 \\ \diagdown O \diagup \\ \vdots \\ HOR \end{array} \qquad (2)$$

The epoxy-oligomer curing with amines has an autocatalytic character due to the accumulation of hydroxyl groups during the reaction [cf. Scheme (1)].

2.2 Noncatalytic Reaction

Until recently, the problem of the general feasibility of a noncatalytic occurrence of this reaction, i.e. via a direct amine-epoxy group interaction, had remained unsolved. A solution of this problem, which is very important for the understanding of the mechanism of nucleophilic addition to the epoxy ring, has become possible only after a thorough purification of the starting material and reaction vessels from the traces of moisture and other hydroxyl-containing compounds [28]. In this way, we succeeded in decreasing hydroxyl-containing impurity concentration expressed as water content down to less than 10^{-6} mol $\cdot$ l^{-1}. This technique has played an important role in the solution of a number of other problems concerning the kinetics and mechanism of epoxy-amine reactions.

Also phenylglycidyl ether (PhGE) reacts with aniline under these conditions [28]. However, the initial reaction rate is proportional to the square of the aniline concentration. The presence of even small amounts of moisture decreases the order of the reaction with respect to amine concentration. The quadratic dependence of the reaction rate on amine concentration A_1 suggests that aniline acts both as an electrophilic substance forming an intermediate complex (A_1E) with the epoxy compound E [cf. Scheme (6)]

$$A_1 + E \underset{K_-}{\overset{K_+}{\rightleftarrows}} (A_1E)$$

and as a nucleophilic reagent (3)

$$(A_1E) + A_1 \xrightarrow{K_1} A_2 + A_1$$

and the addition product A_2 is formed. It is obvious from this Scheme that

$$\left(\frac{d\alpha}{dt}\right)_0 = K_1 \frac{K_+ A_{10}^2}{K_- + K_1 A_{10}}.$$

At $K_- \gg K_1 A_{10}$, since A_1 is a very weak acid, $\left(\dfrac{d\alpha}{dt}\right)_0 = K_1 K_{eq.} \cdot A_{10}^2$,

where $K_{eq.} = \dfrac{K_+}{K_-}$, and $\alpha = \dfrac{E_0 - E}{E_0}$, is the conversion of the epoxy compound.

This equation is in agreement with the experimentally found dependence of the reaction rate on amine concentration. The decrease of the order of the reaction kinetics with respect to amine concentration in the presence of hydroxyl-containing impurities is partially due to the catalytic mechanism when a hydroxyl-containing compound, whose acidity is far higher than that of amine, acts as a proton donor. It should be stressed that, with the usual technique of reagent purification and cleaning of the reaction vessels, the moisture content in the reaction mixture is still rather high, and the order of the reaction kinetics with respect to aniline concentration considerably differs from two.

This result is quite clear in view of the importance of the proton donor compounds and the fact that the observed trimolecular rate constant of the catalytic reaction is higher by a factor of 40–50 than that of the noncatalytic reaction [29, 30].

The parallel occurrence of the two reactions at a low concentration of the proton donors (water, alcohols, phenols, acids) can lead to a very complicated kinetics. Thus, in the investigations of the order of the reaction of PhGE with an aliphatic amine (butylamine) [31] the reaction order with respect to the amine concentration is a fractional one and depends on the reaction temperature; thus, at 12, 28, 42 and 58 °C the values of the order are 1.8, 1.6, 1.4 and 1.2, respectively. In the whole temperature range, the initial rate is well described by a superposition of two reactions, one of which is proportional to the first and the other to the second power of the amine concentration. The contribution of each reaction depends on temperature, for their activation energies are different: $E_1 = 42$ kJ · mol^{-1} and $E_2 = 25$ kJ · mol^{-1}, respectively.

Estimations [32] show that under the experimental conditions selected the concentration of the hydroxyl groups on the glass surface of the reaction apparatus is about 5×10^{-2} mol l^{-1} related to the water content. Thus, the reaction of PhGE with amine proceeding by the first order with respect to amine concentration is considered to be a catalytic one, the epoxy ring being activated by the hydroxyl-containing impurities in the reaction system. This conclusion is confirmed by a comparison of the effective rate constants of the third order for the catalytic (first order with respect to amine concentration) and noncatalytic (second order with respect to amine concentration) reactions of PhGE with butylamine, the ratio of rate constants of which is ~ 40, a value similar to that for aromatic amines [29, 30].

Therefore, the reaction of an epoxy compound with amine can proceed in the absence of hydroxyl-containing compounds specially added or occurring as impurity, the amine acting both as a nucleophilic and an electrophilic reagent.

It is to be noted that there are essentially no principal differences between the catalytic reaction, when the ring is activated by a hydroxyl-containing component, and the noncatalytic one when amine acts as an electrophilic reagent (proton donor). It should be emphasized that for epoxy ring opening by such weakly nucleophilic reagents as amines, an electrophilic assistance is essential. This idea will be confirmed below several times because it is the key idea for understanding the kinetic mechanism of the reaction of epoxy compounds with amines.

Can this conclusion be extended to the other known reactions of nucleophilic addition to the epoxy ring? This question remains so far open and will be discussed below (Sect. 3).

2.3 Addition Mechanism

Let us consider in detail the mechanism of addition of amines to the epoxy ring. The catalytic addition of amines [Scheme (2)] proceeds through a preliminary formation of fairly stable donor-acceptor complexes of the epoxy compound with hydroxyl-containing reagent (cf. Sect. 2.4) with a subsequent nucleophilic attack of the amine on this complex. In other words, this quasi-trimolecular reaction actually proceeds through two successive bimolecular reactions with the realization of the "push-pull" mechanism in the transition state. It should be noted that during formation of the donor-acceptor complexes of the amine with the hydroxyl-containing compound ($A_1 \ldots HOR$) the amine is in fact deactivated as a nucleophilic reagent so that only complex-unbound amine molecules fulfill this role. The detailed mechanism of a noncatalytic reaction of the epoxy compound with amines is not yet clear. In addition to the mechanism similar to that for the catalytic reaction described above [Scheme (3)], one can also imagine an alternative mechanism consisting in the initial formation of a more reactive amine dimer

$$2A_1 \rightleftharpoons \underset{(A_1)_2}{\overset{H^{\delta+}}{\underset{}{>}}N: \cdots H - \overset{\delta-}{\underset{}{\ddot{N}}}<} \tag{4}$$

where according to Scheme 4 both the electrophilic and nucleophilic activities must be higher compared with those of the monoamine. A kinetic scheme in this case may be formulated as follows

$$\begin{aligned} & 2A_1 \rightleftarrows (A_1)_2 \\ & (A_1)_2 + E \rightleftarrows [(A_1)_2 E] \rightarrow A_2 + A_1 \, . \end{aligned} \tag{5}$$

The corresponding transition states for these alternative mechanisms [(3) and (5)] may be represented in the following form

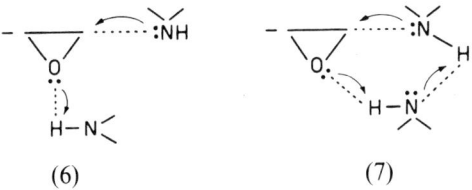

(6)   (7)

According to the available experimental data, it is impossible to distinguish between these mechanisms, but the second mechanism seems to be preferred [Scheme (7)] for, according to this Scheme, the reaction of amine addition proceeding through a cyclic transition state is completed in one step, whereas for the reaction to occur according to Scheme (2) or (6) it is additionally necessary to transfer the proton. Then, it is probable that the different mechanisms [Schemes (3) and (5)] may precede formation of one and the same transition state [Scheme (7)]. Note finally that the mechanism of bifunctional catalysis [Scheme (7)] is extremely popular in different reactions of nucleophilic substitution at the saturated carbon atom and reactions with participation of a carbonyl group [32].

2.4 Reaction Mechanism

The mechanism of addition of amines to the epoxy ring considered above is part of the general problem of the mechanism of the reaction under study. The main feature of this mechanism is formation of different hetero- and autocomplexes of the starting substances and reaction products characterized by different reactivity [5, 13, 14, 33]. These complexes may be formed both inter- and intramolecularly. The existence of these complexes has been proved and the thermodynamic parameters in complexing reactions have been determined [5, 13, 14, 29, 33–43].

The idea of the decisive role of complexing reactions makes possible to unterstand such peculiarities of epoxy compound reactions with primary and secondary amines as a rather unusual, from the kinetic point of view, simultaneous occurrence of the autoacceleration and autoinhibition reactions, catalytic and inhibitive effects of various solvents, and obvious ortho-effect of a number of substituents in aromatic amines, the influence of tertiary amines as additives and some other [5, 6, 13, 14, 30, 44–46].

So, the curing mechanism of epoxy oligomers with amines is rather complicated and, to a first approximation, may be described by Scheme (8) taking into consideration all the donor-acceptor interactions of the starting reagents with the reaction products and with each other

1. $2 A_1 + E \rightarrow A_2 + A_1$
   $2 A_2 + E \rightarrow A_3 + A_1$
2. $A_1 + EH_n \rightarrow A_2 + H_n$
   $A_2 + EH_n \rightarrow A_3 + H_n$
3. $E + H_n \rightleftarrows EH_n$

4. $A_i + H_n \rightleftarrows A_i H_n$ \
5. $H_n + H \rightleftarrows H_{n+1}$ \
6. $A_i + A_j \rightleftarrows A_{i+j}$, \hfill (8)

where $i = 1, 2, 3$; $n = 1, 2, 3 \ldots$; $j = 1, 2$; A_1, A_2, A_3 are the primary, secondary and tertiary amine groups, respectively; H is the free hydroxyl group; H_n are the autocomplexes of the hydroxyl group; A_{i+j} are autocomplexes of the amine groups; E is the epoxy group.

This Scheme is to be supplemented with complexing reactions between all the proton donors in the system and electron donor atoms in the starting components (for example, oxygen atom in the PhO ether bond in a phenylglycidyl ether molecule) and the corresponding reaction products.

Scheme (8) does not include additives of hydroxyl-containing reagents, and H denotes the hydroxyl groups formed as a result of the reaction. If a hydroxyl-containing compound is added to the system, the kinetic Scheme (8) should be supplemented. Finally, in the formation of autocomplexes it is necessary to distinguish linear and cyclic forms whose reactivities are rather different. The cyclic forms are completely deactivated, i.e. they do not take part either in reaction 2 or in reactions 3–5 in Scheme (8).

2.5 Reaction Thermodynamics

Thermodynamics of the epoxy compound curing with amines is interesting from two points of view. It concerns all the numerous elementary reactions resulting in the formation of H-complexes in the reaction system and the process as a whole, as well.

It is impossible to understand all the complicated kinetic paths and to determine the elementary reaction rate constants without a detailed quantitative investigation of all the donor-acceptor interactions in the reaction system. Strictly speaking, at present there are no data on the elementary reaction rate constants even in low-molecular model systems.

Investigation of various donor-acceptor interactions in a polymer, where all the chemical reactions have been completed, is also of great interest because these interactions considerably affect its physico-mechanical properties.

2.5.1 Donor-Acceptor Complexes

Table 1 presents values of thermodynamic parameters of complexing reactions for a number of model systems obtained with the help of calorimetry and IR-spectroscopy [14, 34, 35].

As can be seen, the alcohol autocomplexes possess the highest strength; the bond strength in the series of amine complexes with alcohol falls from the primary to the tertiary ones, which proves the decisive role of steric factors in the formation of H-complexes with amines. At low temperatures, the equilibrium constant of the A_iH complexes falls in the series $K_{eq.1} > K_{eq.2} > K_{eq.3}$; however, at temperatures over 60–70 °C the series becomes reversed. It means that during the reaction at elevated temperatures not only the absolute but also the relative concentrations of the A_2H

Table 1. Thermodynamic parameters for the formation of model donor-acceptor complexes [14]

System	$-\Delta H$ (kJ · mol^{-1})	$-\Delta S$ (cal · mol^{-1} · degr.$^{-1}$)	K_{eq}* Calorim. (l · mol^{-1})	K_{eq}** IR-spectr. (l · mol^{-1})
Cyclohexanol-cyclohexanol (HH)	27.2	98.6	—	—
Cyclohexanol-phenylglycidyl ether (EH)	14.2	46.0	1.34/0.48	1.47/0.6
Cyclohexanol-aniline (A$_1$H)	18.8	67.0	1.85/0.45	2.00/0.5
Cyclohexanol-N-ethylaniline (A$_2$H)	11.7	41.8	1.35/0.73	1.16/0.63
Cyclohexanol-N,N-diethylaniline (A$_3$H)	3.8	8.4	0.63/0.55	0.30/028

* data for 22 °C/97 °C
** data for 28 °C/98 °C

and A$_3$H complexes increase in the epoxy compound-aromatic amine system. This result is most important for the understanding of the autoinhibition effect in similar systems (cf. Section 2.8.1).

From the experimental kinetic data obtained by isothermal and adiabatic calorimetry, a technique for determining the kinetic and thermodynamic parameters for a somewhat simplified Scheme (8) has been developed. Table 2 presents thermodynamic parameters for two models and a real systems.

The table demonstrates a good agreement between the thermodynamic parameters of the complexing reactions inside both the model and the real systems. Moreover, the absolute values of the thermodynamic parameters for the corresponding complex-

Table 2. Thermodynamic parameters of complexing reactions for a series of epoxy amine systems [5, 16, 17, 29, 30]

Complex	$-\Delta H$ (kJ · mol^{-1})	$-\Delta S$ (J · mol^{-1} · degr.$^{-1}$)
Phenylglycidyl ether-aniline [14, 29]		
EH	9.6	29.3
A$_1$H	14.2	35.5
A$_2$H	9.2	27.2
Phenylglycidyl ether-N-ethylaniline [30]		
EH	10.4	33.4
A$_2$H	10.4	41.8
A$_3$H	4.6	5.0
Diglycidyl ether of resorcinol-2,5-diaminopyridine [5, 16, 17]		
EH	16.3	33.4
A$_1$H	14.2	31.3
A$_2$H	10.4	41.8
A$_3$H	4.6	5.0

reactions are almost the same for all systems, and they are close to those for the systems where model alcohol is used (Table 2).

A detailed consideration of the complexing of the hydroxyl groups with all the electron-donor groups in the polymer has been obtained by use of the IR-spectroscopic technique on cured epoxy resins and model compounds [33, 36].

The identified structures of the donor-acceptor complexes in the epoxyamine polymers are presented in Schemes (9) and (10).

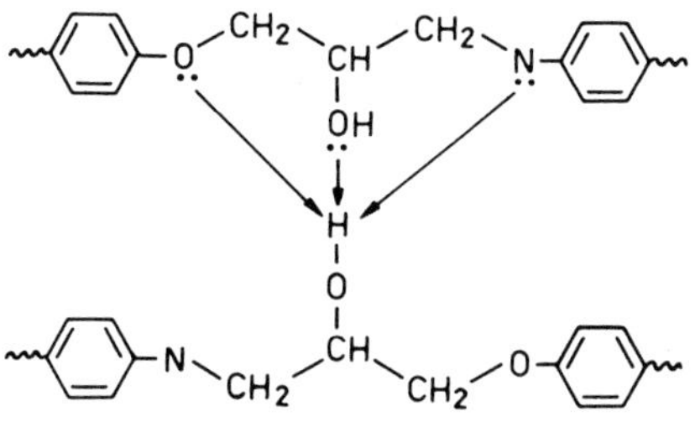

Intermolecular complexes (9)

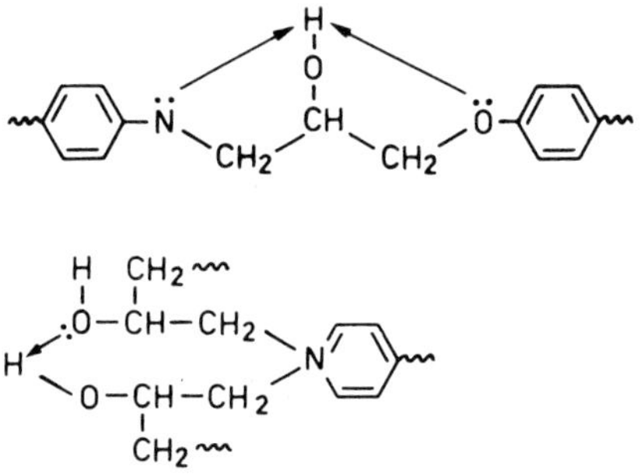

Intramolecular complexes (10)

The existence of the above complexes has been conclusively proved by use of specially synthesized model compounds (Table 3) and by IR-spectroscopic studies over a wide range of substance concentrations (diluent CCl_4).

Table 3. Model compounds [33]

Serial number	Formula
I	$PhOCH_2CHCH_2O$—⟨⟩—OCH_2CHCH_2OPh with OH, OH
II	$PhOCH_2CHCH_2NCH_2CHCH_2OPh$ with OH, Ph, OH

Table 3. (continued)

Serial number	Formula
III	PhNHCH$_2$CH(OH)CH$_2$OPh
IV	C$_4$H$_9$NHCH$_2$CH(OH)CH$_2$OPh
V	Ph–OCH$_2$CH(OH)CH$_2$N(C$_4$H$_9$)CH$_2$CH(OH)CH$_2$OPh
VI	PhN(CH$_3$)CH$_2$CH(OH)CH$_2$OPh
VII	![piperazine-like structure with N=N] with four PhOCH$_2$CH(OH)CH$_2$– groups attached (two PhOCH$_2$CH(OH)CH$_2$– on left N and two –CH$_2$CH(OH)CH$_2$OPh on right N)
VIII	[–N(Ph)CH$_2$CH(OH)CH$_2$O–C$_6$H$_4$–OCH$_2$CH(OH)CH$_2$–]$_n$ n=90

The 5-membered cycles are shown to be characteristic of the cyclic associates [33]. In molecules of the model compounds II and V there also exist 8-membered cycles with OH ... OH hydrogen bonds. However, the total fraction of cyclic associates in the polymer is not great (1–2%). A similar conclusion has been made earlier from kinetic studies [14, 44].

Table 4. Thermodynamic parameters of the model H-complexes [33, 38, 39]

System	Type of H-bond	Equilibrium constant at 298 K (l · mol^{-1})	$-\Delta H$ (kJ · mol^{-1})
1 Isopropanol-isopropanol	OH ... OH	2.45	17.21
2 Isopropanol-phenylglycidyl ether	OH...O (O◁)	1.08	—
3 Isopropanol-anisole	OH ... O	0.33	9.6
*)	OH...O◁	0.75	12.5
4 Isopropanol-N,N-diethylaniline	OH ... N	0.38	9.2
5 N-methylaniline-N-methylaniline	NH ... NH	0.45	7.5
6 N-methylaniline-anisole	NH ... O	0.36	7.5
7 N-methylaniline-di-n-propyl ether	NH ... O	0.29	13.0

*) The values are the difference between K and ΔH of systems 2 and 3

Studies of the model compounds III and IV [33] and others [39] have shown that the NH-groups take part in hydrogen bonding both in the crystalline and amorphous states not only as electron donors but also as electron acceptors. Table 4 presents thermodynamic parameters for a series of model complexes.

As can be seen from Table 4, self-association of the hydroxyl groups is a predominant type of association at room temperature. The fraction of the self-associates of the hydroxyl groups in the epoxy-amine networks of stoichiometric composition amounts to 85–90%. An abundant literature on the self-association of the hydroxyl compounds (alcohols, phenols, acids) [38, 47, 48] shows that a number of different n-mer linear and cyclic self-associates exists in equilibrium. The cyclic associates usually existing in the form of the trimers

$$
\begin{array}{c}
R\diagdown_{O}\diagup^{H\cdots}\diagdown_{O}\diagup R \\
\vdots \quad\quad\quad | \\
H\diagdown_{O}\diagup^{H} \\
|\\
R
\end{array}
\tag{11}
$$

are actually inactive in catalysis of the nucleophilic epoxy-amine addition, whereas the activity of the linear associates increases substantially as compared with the monomeric hydroxyl-containing compound

$$
\overset{\delta^+}{H-O:}\cdots(H-O:)_n\cdots H-\overset{\delta^-}{O:} \quad\quad n=0,1,2\ldots
\tag{12}
$$
$$\; | \quad\quad\quad |\quad\quad\quad\;\;|$$
$$\;\;R \quad\quad\quad R \quad\quad\quad\;\;R$$

Now, the question arises to what extent the thermodynamic information on the low-molecular-weight model compounds can be applied to real epoxy-amine network polymers. A clear answer to this question is given by a direct comparison of the

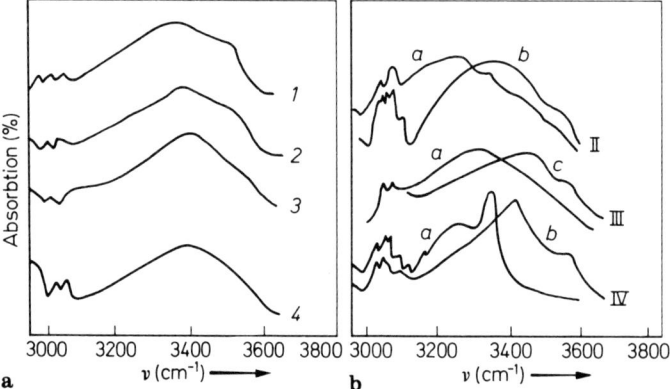

Fig. 1a and b. IR-spectra of epoxy-amine polymers and model compounds at 298 K [36]. **a.** 1 linear polymer VIII (see Table 3), 2–4 stoichiometric networks based on diglycidyl ether of resorcinol and m-phenylenediamine (2) and 4,4′-diaminodiphenylsulfone (3) and diglycidyl ether of hydroquinone and m-phenylenediamine (4). **b.** Model compounds II, III, VI (Table 3), (a) crystal; (b) melt; (c) glass

IR-spectra of model compounds and polymers. As can be seen from Fig. 1, the position of the absorption maxima, line half-widths, relative intensities of the vibration of groups taking part in the hydrogen bonding are practically the same for the model compounds (melt or glass) and glass-like linear and network polymers. It means that the structure and mean energies of the hydrogen bonds of the epoxy-amine systems are also very much the same. A similar conclusion can be made by a direct comparison of the thermodynamic parameters of the model and polymer systems (Tables 1 and 2).

These results indicate that studies of the donor-acceptor interactions on the model systems are quite justified. This study is the only possible approach to quantitative characterization of all the numerous complexes appearing in the epoxy-amine system. Today, we are making but initial efforts in the thermodynamic studies of the epoxy-amine systems. For the time being, we have only managed to estimate the effective thermodynamic characteristics in such systems. The development of an algorithm for both the experimental and the theoretical approach to the study of similar systems still remains an important task.

2.5.2 Change of the Reaction Enthalpy

The second important aspect in thermodynamic studies is the determination of the enthalpy. A knowledge of the thermochemistry of epoxy-amine interactions is important also as a prerequisite for rational curing processes as manufacturing methods. The solution of this problem is also important for the application of the calorimetric method to the kinetic investigations. In fact, in the case of reactions with continuously varying concentrations of the donors and acceptors, the observed heat release (Q) may depend nonlinearly on conversion (α) as of the general case

$$Q = q\alpha + \sum_i q_i(\alpha), \tag{13}$$

where q is the thermal effect of the epoxy ring opening. It is assumed that the value of is equal both for the primary and secondary amines since the major fraction of q is due to the heat of the epoxy cycle opening; $\sum_i q_i(\alpha)$ is the total thermal effect of all the

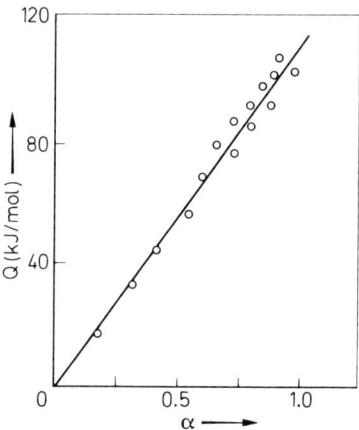

Fig. 2. Dependence of heat release on conversion [29] in the reaction of phenylglycidyl ether with aniline

donor-acceptor interactions in the system depending nonlinearly on the conversion $\alpha = \dfrac{E_0 - E}{E_0}$. Fortunately, the values of Q and α determined by the chemical or IR-spectroscopic methods exhibit a direct proportionality (Fig. 2). It means that $\sum_i q_i(\alpha) \to 0$ and $q\alpha \gg \sum_i q_i(\alpha)$, i.e. in these systems one can observe a compensation of the thermal effects of different complexing and destructive reactions. It implies that the calorimetric method can be applied to the kinetic studies of such reactions.

Table 5 presents reaction enthalpies measured by the calorimetric method for a number of epoxy-amine systems of stoichiometric composition [29, 49-52].

Table 5. Change in the enthalpies of the reactions of the epoxy compounds with primary amines

Epoxy compound	Amine	$-\Delta H$ (kJ · mol^{-1})	Ref.
Phenylglycidyl ether	Aniline	110 ± 2	49)
Phenylglycidyl ether	Aniline	110 ± 2	29)
Phenylglycidyl ether	Butylamine	115 ± 3	49)
Phenylglycidyl ether	Butylamine	102 ± 3	50)
Phenylglycidyl ether	Hexamethylenediamine	116 ± 2	49)
Diglycidyl ether of diglycidyl ether of 4,4'-hydroxyphenyl-2,2-propane (Bisphenol A)	Aniline	118 ± 1	49)
Diglycidyl ether of diglycidyl ether of 4,4'-hydroxyphenyl-2,2-propane (Bisphenol A)	4,4'-Diaminodiphenyl sulphone	112 ± 3	49)
Diglycidyl ether of resorcinol	Aniline	117 ± 2	49)
Diglycidyl ether of resorcinol	4,4'-Diamino-diphenyl sulphone	112 ± 3	49)
Diglycidyl ether of resorcinol	m-Phenylenediamine	101 ± 2	49)
Diglycidyl ether of diglydidyl ether of 4,4'-hydroxyphenyl-2,2-propane (Bisphenol A) oligomers at M = 400–1000	m-Phenylenediamine	101 ± 2	42)
Diglycidyl ether of diglycidyl ether of 4,4'-hydroxyphenyl-2,2-propane (Bisphenol A) oligomers at M = 400–1000	m-Phenylenediamine	109 ± 2	51–53)

As can be seen, the enthalpies of different apoxy-amine systems, according to different authors, lie in a rather narrow range (100–118 kJ per mole of epoxy groups, i.e. close to the heat of the epoxy ring opening). These data confirm the above conclusion as to the small total contribution of the donor-acceptor interactions in the epoxy-amine systems to the observed integrated value of the heat release and the possibility of the application of the isothermal calorimetry method to the reaction kinetic studies.

At the same time, one should pay attention to the fact that in some systems the reaction can be accompanied by phase transition. In this case [49], the effective enthalpy of the reaction is higher by the enthalpy of the phase transition. Thus, the heat of crystallization of the products formed by interaction of phenylglycidyl ether with hexamethylenediamine makes up 42 kJ · mol^{-1}, i.e. the enthalpy of the resulting crystalline, rather than liquid product, will be 158 kJ · mol^{-1} instead of 116 kJ · mol^{-1} [49].

A combined application of direct calorimetric measurements and thermochemical investigations has made possible to obtain a number of important thermochemical quantities characterizing the interaction of the N—H bond of the amine with the epoxy ring [53]. Combustion and evaporation enthalpies of phenylglycidyl ether and its condensation products with aniline and butylamine have been determined. Standard enthalpies of the formation of these compounds, strain energies of the epoxy ring in the phenylglycidyl ether molecule and $-\Delta H$ values for the three-phase states, which are most important for the determination of the true thermodynamic reaction characteristics, have been estimated.

2.6 Reaction Kinetics

2.6.1 General Kinetic Principles

As has been stated above, the autocatalytic course is a characteristic feature of the kinetics of the reaction under study. Generally, the reaction rate is third order with the first order with respect to the concentrations of the reagents, viz., the epoxy compound, amine and proton donor. In the absence of proton donor molecules, the amine molecule itself can play the role of proton donor. Note that the contribution of the latter reaction to the total reaction rate in a later stage of the reaction is negligible. The observed trimolecular reaction rate constant changes with conversion and depends on the concentration of the starting reagents [33, 54], which is quite clear from the view point of the above mechanism. Nevertheless, these changes are often not significant [33] which suggests the presence of a certain compensation mechanism in such systems, which will be considered below.

The kinetic course of the process is much simpler if the reaction takes place in excess of alcohol. In this case, the maximum reaction rate is observed in the very beginning of the reaction and the rate is described by the kinetics of a simple successive bimolecular (actually, quasi-bimolecular) reaction [55]. This procedure has been used by most researchers studying the kinetics of reactions of epoxy compounds with amines [50, 55–63]. Unfortunately, the kinetic parameters obtained by different authors cannot be correlated since they depend on the nature of the alcohol used, exhibiting an increase with its acidity [55, 56]. On the other hand, the reaction rate constants obtained by using this approach are expected to depend on alcohol concentration and their values vary considerably. Nevertheless, a comparative study of the quasi-bimolecular rate constants under the same experimental conditions may serve for comparison.

The trimolecular rate constants were determined in earlier works with no account of any donor-acceptor interactions in the system [26, 50], but later on an attempt has been made to consider such reactions [5, 14, 16, 17, 29, 30]. The effective kinetic and thermodynamic parameters that are independent of conversion are useful in describing the kinetic curve (such as that in Fig. 5), although the interpretation of the physical meaning of these constants is still very tentative. Such a situation seems to be typical of all liquid-phase reactions in which an important role is played by the various auto- and heteroassociates of the components.

2.6.2 Structure of the Effective Reaction Rate Constant

Let us consider this problem in more detail. In fact, the autoassociates of the hydroxyl-containing compounds and amines can be composed of molecules, having different reactivities in the interaction with amine. Therefore, the actually observed reaction rate constant is complex in its composition. Thus, even the simplest noncatalytic (in the absence of proton donors) reactions of the epoxy compound with amine, considering all the donor-acceptor interactions, is generally described by the following kinetic scheme:

$$
\begin{aligned}
&1.\ A_i^l + A \overset{K'_{eqi}}{\rightleftarrows} A_{(i+1)}^l \\
&2.\ A_i^l + E \overset{K''_{eqi}}{\rightleftarrows} (A_i^l E) \\
&3.\ A_i^l \overset{K'''_{eqi}}{\rightleftarrows} A_i^c \\
&4.\ (A_i^l E) + A_j^l \overset{k_{ij}}{\longrightarrow} \text{product} + A_{(i+j-1)}^l,
\end{aligned}
\tag{14}
$$

where A is the primary or secondary amine and the superscript l or c denotes the linear (reactive) or the cyclic (nonreactive) autoassociate, respectively. For reaction 1, 2, 4, i = 1, 2, 3 ...; for reaction 3, i = 2, 3 ..., j = 1, 2, 3 ... K'_{eqi}, K''_{eqi}, K'''_{eqi} and k_{ij} are the equilibrium constants and the effective rate constant, respectively. For glycidyl ethers and an electron-donor solvent, Scheme (14) should be supplemented by reactions of amine complexing with an out-of-ring oxygen atom in the ether molecule and with the electron-donor atom of the solvent molecule. If this mechanism is operative, the amine acts as a proton-donor:

$$\text{>N–H} \cdots : \text{O<}. \tag{15}$$

The experimentally observed noncatalytic reaction rate constant k_1 can be calculated as follows

$$W_0 = k_1 E_0 A_{10}^2, \tag{16}$$

where W_0 is the initial rate of the reaction and the subscript 0 corresponds to parameters at t = 0. From Scheme (14)

$$W_0 = \sum_{i=1}^{\infty} \sum_{j=1}^{\infty} k_{ij} (A_i^l E) A_j^l, \tag{17}$$

where $A_i^l E$ and A_j^l are the equilibrium concentrations of the corresponding complexes. At present, it is impossible to solve this equation, i.e. to find the structure of the effective rate constant in the general form.

The real situation is certainly simpler, as the concentration of all complexes shows a sharp drop with the degree of autoassociation. The dependence $k_{ij} = f(ij)$ may be nonmonotonous since the epoxy ring opening probably occurs via a cyclic transition state depending on i and j.

At present, the problem of the structure of the effective rate constant or, in other words, the nature of the transition state in the epoxy ring opening still remains to be solved. As a first approximation, one can neglect the process of amine autoassociation. In this case, the structure of the effective noncatalytic reaction rate constant, assuming that $A_1E \ll E_0$, is clear [cf. Scheme (3)]

$$k_{eff} = kK_{eq} \tag{18}$$

where k is the rate constant of the elementary reaction of the epoxy ring opening [cf. Scheme (7)], K_{eq} is the equilibrium constant of the reaction $A_1 + E \rightleftarrows A_1E$.

A similar structure of k_{eff} will be observed if we assume that the reaction proceeds by Scheme (5) through the same cyclic transition state 7. Here, K_{eq} is the equilibrium constant of the reaction $2A_1 \rightleftarrows (A_1)_2^1$, $A_{10} \gg 2(A_1)_2^1$, the cyclic forms of dimers being absent.

The above consideration indicates that at present, the researchers face large difficulties in the physical interpretation of the effective reaction rate constant associated with the absence of detailed quantitative information on the thermodynamics of formation of H-complexes and their reactivities. Further studies of the donor-acceptor interactions in the epoxy-amine systems will shed more light on this problem.

2.6.3 Compensation Mechanism

Closely related to the problem of the structure of the effective rate constant is the above-mentioned problem of the compensation mechanism. Without a knowledge of this mechanism, it would be impossible to understand why in such a complicated epoxy-amine system one can frequently observe relatively simple kinetic principles, viz., a weak dependence of the effective rate constant on conversion, simple dependences of the initial rate on reagent concentrations, a linear dependence of the total heat release on conversion and almost equal values of the heat release and enthalpy of the epoxy ring opening. The latter two aspects have been discussed above, whereas the first two problems can be understood, say, from a consideration of a noncatalytic reaction.

According to Scheme (14), the initial rate constant is lower as compared with that in the absence of any donor-acceptor interactions due to a decrease in the total number of nucleophilic particles as a result of amine autodissociation. It is higher due to the more pronounced nucleophilic behaviour of that portion of the associated amino groups in which the amine acts as proton donor, as well as due to a certain growth of the acidity of the amine autocomplexes as compared with that of the nondissociated amine molecule. Such a combination of the positive and negative factors leads to a certain levelling off of the role of the donor-acceptor interactions, i.e. to the compensation effect which formally suggests the existence of a linear dependence of the concentrations of the reagents and of their free uncomplexed forms.

2.6.4 Substitution Effect

A key problem in the kinetics of the reactions under study is a relation between the rate constants of the epoxy ring opening under the effect of the primary and secondary amino groups, i.e. manifestation of the substitution effect. This problem has been briefly reviewed by Dušek [64]. Knowledge of the relationship between these rate constants is very important for an adequate description of the kinetics of network formation. It should be emphasized that knowledge of such a relation, rather than the absolute values of the rate constants would be sufficient.

The published data [5, 14, 16, 17, 29, 30, 50, 54–64] on the relation between the constants of the addition of primary and secondary amines are most discrepant. On one hand, such a situation is due to the different experimental conditions, and on the other hand, to the dissimilar methods of calculating the kinetic constants. Most researchers used the kinetic data for reactions in excess alcohol to estimate the quasibimolecular rate constants of the primary and secondary additions. The constants were estimated assuming the validity of a simple scheme of the successive bimolecular reaction [Scheme (1)].

As has been stated above, in such experiments the estimated rate constants depend on the concentration of the starting reagents and the nature of the alcohol used [33, 55, 56]. Therefore, the k_2/k_1 ratios are valid under the assumption of the equal effect of the two factors on interactions of the primary and secondary amine.

Another series of works [60, 61] directly estimated the constant ratios by performing the reactions without any solvent through variation of the initial ratios of the amino and epoxy groups and analyzing the unreacted components and reaction products by the chemical method or by gel-permeation and liquid chromatography. The reactions were performed to full conversion of epoxy groups using an excess of amine to prevent epoxy homopolymerization. The evaluation of the ratio of rate constants is based on the assumption that the donor-acceptor interactions affect k_1 and k_2 in the same way or, in other words, the ratio of catalytic and noncatalytic rate constants entering k_1 and k_2 is assumed to be independent of conversion. If it is not so, the results are affected if low-conversion data are used for calculation of the ratio.

Finally, a series of works [5, 14, 16, 17, 29, 30] attempted to estimate the effective kinetic parameters of both the noncatalytic and catalytic reactions by taking into account all the donor-acceptor interactions [Scheme (8)] except for autoassociation reactions.

The principal possibility of estimating the rate constant of the reaction under study from the experimental measurements of the various parameters describing the struc-

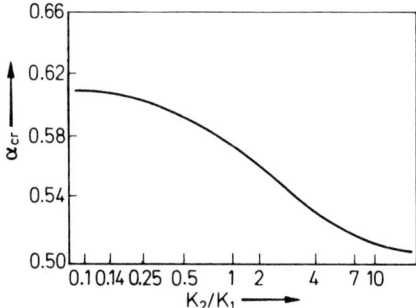

Fig. 3. Dependence of the critical gelation on the k_2/k_1 ratio in the reaction of diamines with diepoxy compounds [3]

ture of the resulting network polymer, such as a critical gelation conversion, sol fraction, equilibrium modulus of elasticity, etc. [64], should be neglected because of a relatively weak sensitivity of these parameters to the ratio of the reaction rate constants. Thus, as can be seen from Fig. 3 [3], the critical gelation for the diepoxide-diamine reaction even in the region of the highest sensitivity changes by ca. 10% as the relative rate constant changes by two orders of magnitude.

Table 6 summarizes the kinetic characteristics of the primary and secondary additions for a number of epoxy-amine systems.

Table 6. Kinetic parameters for a number of epoxyamine systems

System	Reaction conditions	$k_1 \times 10^4$	$k_2 \times 10^4$	k_2/k_1	$E_1^*)$ (kJ·mol^{-1})	Ref.
1 TGE + p-methoxyaniline	Ethanol, 333 K	5.0	2.0	0.4	—	[57]
2 TGE + p-p-toluidine	Ethanol, 333 K	4.1	1.0	0.24	—	[57]
3 TGE-p-chloraniline	Ethanol, 333 K	0.	—	0.24	—	[57]
4 TGE + p-bromaniline	Ethanol, 333 K	0.71	0.18	0.25	—	[57]
5 TGE + p-iodaniline	Ethanol, 333 K	0.64	0.19	0.30	—	[57]
6 TGE + aniline	Ethanol, 333 K	2.0	0.50	0.25	—	[57]
7 TGE + p-cyananiline	Ethanol, 333 K	0.07	0.02	0.24	—	[57]
8 TGE + p-nitroaniline	Ethanol, 333 K	0.04	0.03	0.62	—	[57]
9 IBGE + n-butylamine	Tert-butanol, 323 K	2.85	1.0	0.36	64 / 52	[55]
10 TGE + 4,4'-diaminodiphenylmethane	Ethanol, $E_0 = 0.1$ to 0.4 mol·l^{-1}, 333 K	7.3	2.2	0.33	—	[58]
11 TGE + p-phenylenediamine	Ethanol, $E_0 = 0.1$ to 0.4 mol·l^{-1}, 333 K	25.2	8.9	0.35	—	[58]
12 TGE + benzidine	Ethanol, $E_0 = 0.1$ to 0.4 mol·l^{-1}, 333 K	6.4	1.95	0.31	—	[58]
13 TGE + 4,4'-diaminodiphenylsulphone	Ethanol, $E_0 = 0.1$ to 0.4 mol·l^{-1}, 333 K	0.21	0.11	0.52	—	[58]
14 PhGE + n-dodecylamine	Bulk, 323 K	—	—	0.41	—	[60]
15 DGEBA + n-dodecylamine	Bulk, 323 K	—	—	0.41	—	[61]

System	Reaction conditions	$k_1 \times 10^4$	$k_2 \times 10^4$	k_2/k_1	E_1*) (kJ·mol^{-1})	Ref.
16 DGEBA + 1,6-hexamethylenediamine	Bulk, 323 K	—	—	0.33	—	61)
17 DGEBA + 4,4'-diaminodiphenylmethane	Bulk, 353 K	—	—	0.2	—	61)
18 DGER + aniline	Bulk, 343 K	—	—	0.5	—	63)
19 DGER + phenylenediamine	Bulk, 343 K	—	—	1.0	—	63)
20 PhGE + m-phenylenediamine	Bulk, 343 K	—	—	1.0	—	63)
21 PhGE + n-butylamine	Bulk, 343 K	0.45	0.29	0.65	54–59	50)
22 PhGE + aniline	Bulk, 373 K	—	—	0.44	—	54)
23 PhGE + aniline	Bulk, 363 K	$\frac{0.05}{2.20}$	—	$\frac{0.6}{0.25}$	$\frac{46}{38}$	14, 29)
24 PhGE + N-ethylaniline	Bulk 363 K	—	$\frac{0.003}{0.58}$	—	$\frac{56}{41}$	14, 30)
25 DGER + 2,6-diaminopyridine	Bulk, 360 K	$\frac{0.002}{0.15}$	0.17	1.12	$\frac{42}{42}$	16, 17)

TGE = p-tolylglycidyl ether, PhGE = phenylglycidyl ether, IBGE = isobutylglycidyl ether, DGEBA = diglycidyl ether of bisphenol A, DGER = diglycidyl ether of resorcinol. *) E_1 = mean activation energy or E_1/E_2. For systems 1–20, bimolecular and for systems 12–25 trimolecular rate constants are given (in l·mol^{-1}·s^{-1} and l^2·mol^{-2}·s^{-1}, respectively). For systems 23–25, the values of k_1 and k_2 refer to noncatalytic/catalytic rate constants. E for system 25: E'_1/E'_2. For systems 1–13, 14–20 and 23–25 the rate constant ratios were calculated from the kinetics with excess of alcohol, using GPC separation and taking into account the donor-acceptor complexes, respectively. All the rate constants and their ratios are expressed per one reacting group.

It is evident from Table 6 that the rate constants differ substantially depending on the method of determination. One can observe a fairly distinct negative substitution effect, which suggests a somewhat stronger steric rather than inductive effect. Only for the system 25, this effect is slightly positive. Irrespective of the method of determination, the relative constants for the various systems differ insignificantly and are almost independent of temperature. The rate constant of the noncatalytic reaction (Systems 23 and 24 in Table 6) for which k_1 and k_2 differ by over one order of magnitude is an exeption. However, it is evident that the role of the noncatalytic reaction as compared with that of the catalytic one in the network formation is negligible.

The observed variations in the reactivities of the primary and secondary amino groups result in a shift of the gel point in the diepoxide-diamine reaction towards higher conversions by 4% at the most (Fig. 3). If the substitution effect is not operative, the critical gel point is equal to 0.577 (Fig. 3). The remaining structural characteristics are still less sensitive to the changes in reactivity of the amino group during the reaction.

Thus, the analysis of the reactivity ratios of the primary and secondary amino groups indicates that for conventional curing agents this cannot be regarded as a serious factor affecting the network topology.

Another aspect of the substitution problem consists in determining the reactivity of the second amino or epoxy group in diamines or diepoxides, respectively, after the first group has already reacted.

As reported in numerous studies [61–66], the reactivity of the glycidyl groups in Bisphenol A diglycidyl ether is virtually the same. At the same time, a small negative [54, 67] or positive [5] substitution effect has been observed. However, in some other cases this effect can be substantial [20].

In diamines such as 4,4'-diaminodiphenylmethane, 4,4'-diaminodiphenyl sulphone or 1,6-hexamethylenediamine, the substitution effect is practically absent [64] although this may not be the case for di- and polyamines where the distance between the amino groups is not large allowing inductive and steric effects to be operative [59, 68].

It can be concluded that at present the substitution effect in epoxy-amine reactions is not completely understood, so that additional studies are required.

2.7 Kinetic Feature of the Deep Conversion Stages

2.7.1 Autoinhibition Effect

It is well known that on curing epoxy oligomers with amines, the reaction frequently almost completely terminates long before the functional groups of one of the reagents have been exhausted, which is usually attributed to the diffusion hindrances appearing in the solid polymeric matrix (see Sect. 2.7.2). However, kinetic studies of the PhGE-aniline interaction in o-dichlorobenzene solution indicate that a similar effect, i.e. a sharp inhibition of the reaction long before the starting reagents have been exhausted, takes place too [44], when only low-molecular substances are formed. Nonetheless, a complete conversion of the reacting groups can be reached. The reason for this phenomenon should be sought not in the changes of the physical properties of the environment and transition of the reaction into the region of diffusion control, but rather in certain chemical processes responsible for the inhibition of the reaction.

It has been shown [44] that the inhibition effect observed at high conversions is due

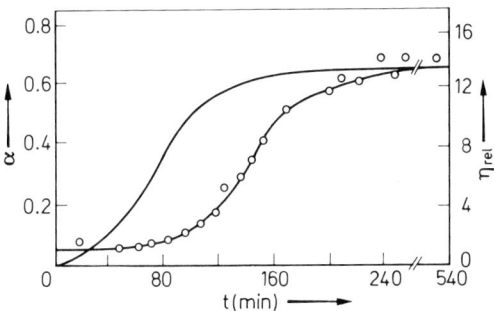

Fig. 4. Kinetics of phenylglycidyl ether consumption (1) and relative viscosity changes (2) in the reaction of phenylglycidyl ether ($E_0 = 4.4$ mol l^{-1}) with aniline ($A_0 = 4.4$ mol l^{-1}) at 363 K [44]

to formation of the nonreactive complexes A_1H_n [reaction 4 in Scheme (8)]. The complexing lowers the concentration of the free hydroxyl groups and of the reactive complex EH_n, but decreases the concentration of the free amine and thereby inhibits the entire process. It should be noted that inhibition occurs only when the basicity of the appearing secondary or tertiary amino groups is higher than that of the starting primary group. Only then does the concentration of the hydroxyl groups decrease gradually and so does the rate of the process as a whole [13, 14, 44] (see Sect. 2.5.1).

The complexing of the amino and hydroxyl groups distinctly manifests itself through a noticeable viscosity increase at high conversions (Fig. 4) due to network formation with labile hydrogen bonds located in the branching sites since formation of the tertiary amino groups results in the appearance of a three-functional reagent in the system (a tertiary amino group plus two hydroxyl groups):

$$\underset{\underset{\underset{-\overset{|}{N}-}{\overset{\vdots}{C}}}{\overset{|}{OH}}}{PhOCH_2\overset{|}{C}HCH_2}\overset{\overset{HO-}{\overset{\vdots}{C}}}{N}\underset{\underset{\underset{-\overset{|}{N}-}{\overset{\vdots}{C}}}{\overset{|}{OH}}}{CH_2\overset{|}{C}HCH_2OPh} \qquad (19)$$

As expected, the effect of the increased viscosity during the reaction becomes more pronounced with decreasing A_{1_0}/E_{U} ratio, where the subscript 0 refers to the reagent concentrations at $t = 0$; the limiting viscosity also increases. At $A_{1_0}/E_0 = 1.8$, 1.0, 0.45, $\lim_{t\to\infty} \eta_{rel} = 3.5, 14.0, 32.5$, respectively. Addition of a stronger base to the reacted system, leads to a sharp drop in the viscosity as a result of the dissociation of the polyfunctional associates.

Increasing reaction temperature results in a smaller fraction of amine to become associated with the hydroxyl groups and, hence, in an increase in conversion at which the process becomes inhibited [29]. The reaction at adiabatic conditions gives a 100% conversion of the epoxide groups without any autoinhibition effect involved [29].

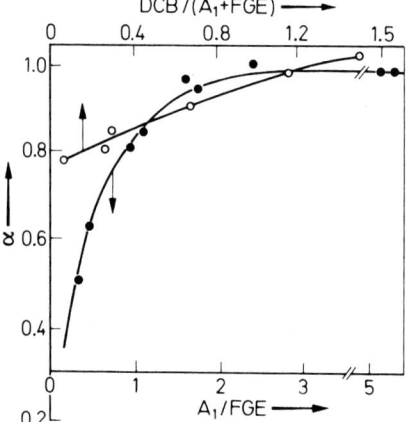

Fig. 5. Dependence of limiting conversion of phenylglycidyl ether on the ratio of reagents (1) and dilution with o-dichlorobenzene (2) in the interaction with aniline (A) at 363 K [43]

As could be expected, the conversion resulting in the reaction inhibition depends on the reagent ratio in that a higher A_{10}/E_0 ratio results in an increase in the limiting conversion (Fig. 5, Curve 1).

Thus, the hydroxyl group fulfills a dual role: (1) due to formation of the EH_n complex it significantly activates the epoxy ring and, (2) upon formation of the non-reactive A_iH_n complexes, the reaction rate decreases as a result of a drop in the concentration of the primary and secondary amines and free hydroxyl groups not entering the A_iH_n complex which catalyzes the reaction. The latter factor has a greater effect on the rate drop than the former, which is confirmed by a direct experiment [14, 44]. Addition of a model secondary alcohol, e.g. cyclohexanol, increases not only the reaction rate but also the conversion at which the reaction becomes be sharply inhibited.

Experiments in diluting the reaction system with o-dichlorobenzene at constant molar ratio of the reagents (Fig. 5, Curve 2) indicate that the conversion at the onset of a sharp drop in the reaction rate increases as the system is diluted. This result suggests a predominant contribution of the intermolecular association and is in good agreement with the rate of change in the relative viscosity [14, 44] and with IR-spectroscopic studies of the nature of the associates [36].

Thus the above data point to the fact that in such systems the reaction mechanism of the epoxy compounds with amines involves not only autocatalysis, but also autoinhibition of the reaction in its deep stages. However, this effect can only be observed if the relative concentration of the free hydroxyl groups decreases due to an increase in hydroxyl-amine complexing as a result of the conversion of the primary into the secondary and then to the tertiary amino group. For the real epoxy-amine compositions, the diffusion mechanism of the reaction inhibition at the deep stages is still the most typical.

2.7.2 Diffusion Control of the Curing Reaction

If the curing reaction occurs at adiabatic or isothermal conditions above the glass transition temperature of the ultimately cured polymer, the reaction kinetics is adequately described by the adopted mechanism [5, 16, 17, 69–71]. However, at temperatures substantially below the final glass transition temperature, the reaction rate at a certain

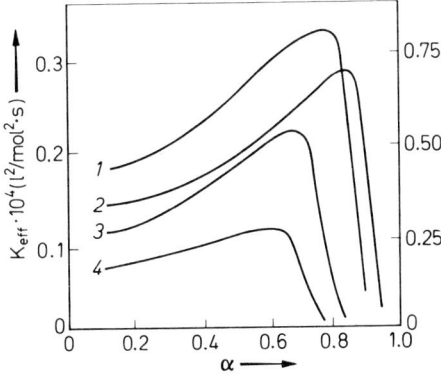

Fig. 6. Dependence of the effective value of the trimolecular rate constant for the reaction of diglycidyl ether of resorcinol with 2,6-diaminopyridine on conversion at different temperatures [79]. (1) 323 K; (2) 333 K; (3) 343 K; (4) 353 K

conversion shows a sharp drop and the reaction completely stops [2, 3, 5, 50, 70–79]. The change of the effective trimolecular rate constant in the curing of diglycidyl ether of resorcinol with m-phenylenediamine at various temperatures (Fig. 6) may serve as an example.

Such a behaviour of the system is attributed to the phenomenon of an isothermal conversion-dependent glass transition. At T_g the conversion reaches the value α_T and then the reaction continues in the glass transition region until the conversion reaches the final value $\alpha_{T,\infty}$. Thus, in the glass-like state $\alpha_{T,\infty} - \alpha_T$, i.e. the polymer with the $\alpha_{T,\infty}$ conversion, obtained at the temperature T, is matched with $T_g > T$ (usually by 15–30 °C). These facts suggest that upon freezing-out the segmental mobility after α_T has been attained as a result of the isothermal concentration-dependent glass transition, the reaction of the functional groups becomes localized so that the reaction rate in the glass-like state will be controlled not by the total concentration of the unreacted groups, but rather by the concentration of the reacting groups alone which, to the instant of the system's glass transition, become adjacent to each other [3, 70, 71]. It is obvious that the lower the glass transition temperature, the higher this concentration with the result that a fraction of the polymer appearing in the glass-like state $\left(\dfrac{\alpha_{T,\infty} - \alpha_T}{\alpha_{T,\infty}} \right)$ increases with lowering curing temperature (see Fig. 6).

The curing reaction starts again when the temperature is raised so that the reaction can be completed by gradually increasing the temperature (Fig. 7).

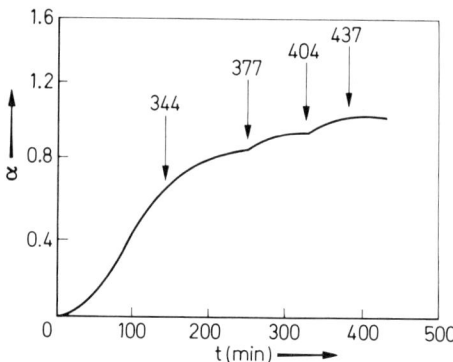

Fig. 7. Kinetics of curing diglycidyl ether of resorcinol with 2,6-diaminopyridine in the stepwise temperature control [3]

The mere fact that the reaction proceeds in the glass-like state implies that for the chemical reaction to occur a small-scale local mobility (possibly β-relaxation) is sufficient, whereas the segmental mobility provides for a translational diffusion of the functional groups in the system [3, 70, 71].

The reaction of curing the epoxy-amine system occurring in the diffusion-controlled mode has little or no effect on the topological structure of the polymer [74] and on its properties in the rubbery state. However, the diffusion control has an effect on the properties of glassy polymers [76–78].

The above mechanism of the diffusion-controlled curing provides a good explanation of the established relation between various physico-mechanical properties of the epoxy-amine polymers in the glassy state and the temperature-time conditions of their preparation [70-71]. The fraction of the strongly nonequilibrated polymer formed in the glassy state depends on the curing temperature which is reflected in the structure and properties. As could be expected, the memory effect of the temperature-time prehistory of the polymers disappears upon annealing in the rubbery state [78]; in other words, a relaxation of the strongly nonequilibrated structures frozen at the curing temperature occurs. Note also that practically no external influence applied to the formed polymer can give rise to the nonequilibrated structures and properties typical of glassy crosslinked polymers.

2.7.3 Bimolecular Reaction Rate Constant for the End Functional Groups During Network Formation

As follows from the preceding paragraph, the system can pass into the diffusion-controlled mode in the course of curing due to a loss of the translational mobility as a result of isothermal conversion-dependent glass transition. This phenomenon is characteristic to the formation of a variety of polymer networks. In the absence of glass transition such slow processes fail to pass into the diffusion-controlled mode during the curing reaction [64]. Nevertheless, recent research [80-86] on the specific features of the relaxation processes in the polymer networks suggests that in principle such a situation is feasible. The influence of these processes on the effective rate constant of the bimolecular chemical reaction can be understood by using the model proposed elsewhere [80].

The space contains arbitrarily fixed ends of two chains of the length L whose free ends carry the functional groups A and B. These groups perform Brownian motion about the fixing site, and upon meeting each other they are capable of reacting. In such a model, the effective bimolecular rate constant is expected to be a function of the distance R between the fixing sites and L. If the chain is shaped as the Gaussian coil, the effective rate constant k(R) is given by:

$$k(R) = k[\beta^3/(2\pi)^{3/2}] \exp(-\beta^2 R^2/2). \tag{20}$$

where $\beta^2 = 3/2L^2$.

It follows from this equation that the constant depends strongly on the ratio R/L and approaches zero at large R/L. Thus, the above model implies that even in the kinetic region the rigid fixing of one of the chain ends carrying the functional groups results in retardation of the reaction at the free end. This effect can be obviously regarded as having a purely topological nature.

2.7.4 Topological Reaction Limit

Essentially the same result has been obtained in studies of formation of an epoxy-amine network by computer-assisted modelling [81]. The imposed absence of any translational mobility of the cross-link sites brings about a drastic drop of the effective rate constant at large conversions. The progressively increasing topological complexity of the system ultimately results in a topological reaction limit excluding the complete conversion of functional groups. This limit is attributed to the fact that the reactive groups in various topological formations cannot meet and react even in the complete

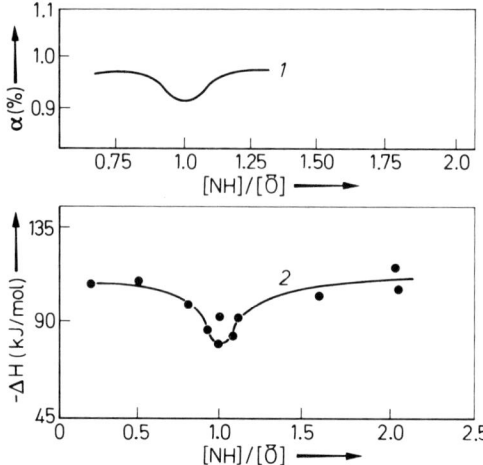

Fig. 8. Dependence of limiting conversion on the ratio of the components: (1) model calculations; (2) diglycidyl ether of resorcinol with m-phenylenediamine [79]

absence of any diffusion hindrances. This theoretical prediction is in good agreement with experiment (Fig. 8). The topological limit becomes particularly evident at the equimolecular ratio of functional groups, whereas at a marked excess of one of the reagents, no topological effect can be observed [79].

It should be pointed out that the topological limit cannot be removed by a simple temperature rise as is the case in reactions controlled by glass transition.

The above models describe a simplified situation of stationary fixed chain ends. On the other hand, the characteristic rearrangement times of the chain carrying functional groups are smaller than the duration of the chemical reaction. Actually, in the rubbery state the network sites are characterized by a low but finite molecular mobility, i.e. R in Eq. (20) and, hence, the effective bimolecular rate constant is a function of the relaxation time of the network sites. On the other hand, the movement of the free chain end is limited and depends on the crosslinking density [82-84]. An approach to the solution of this problem has been outlined elsewhere by use of computer-assisted modelling [85] Analytical estimation of the diffusion factor contribution to the reaction rate constant of the functional groups indicates that $K \sim 1/\tau$, where τ is the characteristic diffusion time of the terminal functional groups [86].

Note finally that the above modifications of the described models [80-81] do not radically alter the former conclusions.

2.7.5 Reaction Kinetics and Polymer Structure

The problem of the reaction kinetics and structure of the resulting polymer has many facets and at present it is far from being solved [2-6, 15, 61, 64, 65, 70, 71, 74-78, 80, 87-97]. It should be noted that the epoxy-amine systems turned out to be the most convenient for experimental and theoretical studies of the process of formation of the topological structure of networks. In many cases their topology in the rubbery state agrees with the theoretical predictions [61, 64, 80, 87, 88].

The computer-assisted modelling of formation of the epoxy-amine networks indicates that the density fluctuations of the crosslinks does not exceed the value predicted by the statistical theory, which is much lower than the size of the globules in amorphous polymers [87, 88]. These facts suggest that the inhomogeneities of the network polymer structure cannot be regarded as responsible for the differing densities of the glassy epoxy-amine polymers observed by various methods.

Earlier, the same conclusion was made by Dušek et al. [92], who believed that the systems under study did not exhibit any special features characteristic of structural inhomogeneities and that their structure does not differ essentially from that observed in the glassy linear polymers. However, a feature that is typical of just the epoxy-amine polymers may exist; it is connected with the autocatalytic formation of epoxy-amine polymers.

Such a possibility follows from a recent study conducted by the present author by use of computer-assisted modelling of the kinetics of autocatalytic reactions occurring under conditions of limited mobility of the reaction products [70, 71, 96].

From this study it follows that such systems are characterized by a pronounced kinetic effect of the reaction inhibition and the resulting product is composed of clusters distributed in the reaction system non-homogeneously in space. The more autocatalytic the reaction is and the lower the mobility of the products, the stronger these two affects are.

It is more likely that under autocatalytic conditions the reaction starts at the site where the product has already been formed. After the reaction products have become less mobile, the reaction is localized in those sites where the products — as molecular clusters — have been formed. It is obvious that the lower the molecular mobility and the stronger the autocatalytic factor, the larger is the size of the growing clusters. In the limiting case, the reaction will proceed similarly to what is observed for the crystallization process [90, 97]. The kinetic inhibition means that the catalysis involves only those molecules that are disposed on the cluster surface, whereas those inside the cluster are essentially inaccessible. Therefore, the larger the cluster size, the higher is the total autocatalysis efficiency. The specified inhibition mechanism is also likely to contribute to the reaction inhibition observed at deep stages of the epoxy-amine reaction (see preceding sections). The most important prediction following from the model (reaction autoacceleration in the case of the limited mobility of the reagents) is the possibility of formation of inhomogeneous structures (supermolecular and phase formations) in such systems. This mechanism is especially pronounced in formation of network polyether acrylates [97]. Revealing the role of this mechanism in the formation of the epoxy-amine system generates a need for further studies.

2.8 Curing of Epoxy-amine Systems at Nonisothermal Conditions

In curing of epoxy-amine compositions, especially in casting large-size thick-wall products, the process — due to a high thermal effect and low thermal conductivity — frequently passes into the nonisothermal mode (quasiadiabatic, or reaction front propagation) so that it becomes practically uncontrollable.

The reaction can be truly adiabatic for a wide range of experimental conditions. It initiates in the kinetic region and reaches a limit of 94–96% conversion, which corresponds to the topological limit for these systems [5, 14, 16, 17, 51, 69]. The reaction kinetics is well described by the adopted mechanism [51, 69], and the resulting experimental kinetic and thermodynamic reaction parameters are similar to those obtained in isothermal experiments.

A similar kinetic description is also valid for the reaction front propagation conditions, although in this case the kinetic equations should be complemented with the thermal conduction equation [5, 16, 17, 98–100]. The curing process under these conditions has been experimentally studied for a number of epoxy-amine systems. These systems are interesting in that under the steady-state conditions the temperature profile precedes the conversion profile, i.e. in the steady-state propagation of the reaction front the latter has a planar shape so that the major portion of the product is formed in a narrow temperature interval at a temperature approximating the adiabatic exotherm. This fact enabled to study the reaction kinetics under the steady-state reaction front propagation conditions by the travelling wave technique [99].

These studies have become a basis for the development of a method of optimizing the curing conditions for large-size thick-wall products made of composite materials with epoxy-amine binders. The curing process tends to occur in inhomogeneous temperature-conversion fields. This leads to residual stresses and even cracks appearing in the products. Using a hypothesis that the residual stresses in the product decrease with decreasing inhomogeneity of the temperature-conversion field in the course of curing and cooling, it has been possible to propose the so-called "thermal theory of the technological monolithic character of designs" [5, 16–18] permitting a calculation of such short-term curing and cooling conditions for the preparation of solid products. Experiments have fully supported the conclusions of this theory [16–18].

2.9 Reagent Structure and Reactivity

There exist only a few systematic studies of the effect of the structure of epoxy compounds and amines on their reactivity during copolyaddition [5, 14, 42, 45, 57, 102].

2.9.1 Amines

Fundamental results elucidating the relationship between the amine structure and reactivity have been obtained in the reaction of various aniline derivatives with p-tolylglycidyl ether in ethanol solution [57]. As could be expected, the reaction rate increases with the basicity of the reacting amine (cf. Table 6), and a good correlation between the logarithm of the rate constant and basicity constant of the amine. The data are well described by the Hammett equation with $\varrho = 1.32$ and 1.27 for reactions of the primary and secondary (substituted) amino group, respectively. The reactivity is raised by electron-donor substituents in the aromatic ring and lowered by electron-acceptor groups [57]. However, some factors indicate an abnormally low reactivity of aromatic amines containing electron-donor groups in the ortho-position. Thus, the reactivity of the amino group with the SCH_3 substituent in o-position is higher by a factor of 20 compared with the unsubstituted compound [102]. A similar picture is observed for amines with methoxy groups in this position [45]

The reactivity of diamine (21) in the reaction with diglycidyl ether of resorcinol is higher by a factor of 10 than that of diamine (22). The two amines with the electron-donor substituents in o-position reflect the existing intramolecular hydrogen bonds where the amino group acts as a proton-donor. It could be expected that the nucleophilic capacity of such an amino group should increase as a result of both the inductive effect of the electron-donor substituent and formation of a donor-acceptor complex [cf. Scheme (4) and Table 4]. The rate of the reaction of this group with α-oxide is expected to increase, but in fact it decreases. Such a contradiction can be readily reconciled if one takes into account that the association involves binding of the amino group hydrogen atoms the concentration of which decreases in the course of reaction. It is clear that according to the existing conceptions of the necessary electrophilic assistance for the nucleophilic attack on the epoxy ring to occur, such a decrease in the proton donor concentration in pure systems, when at the initial stages the amine itself acts as a proton donor, should lead to a decreasing reaction rate which is really the case.

These conceptions were useful in synthesis of a variety of aliphatic and mixed (aliphatic-aromatic) liquid diamines with a lower reactivity [103]

$$A_r NHCH_2CH_2(OCH_2CH_2)_m NH_2 \quad m = 1, 2$$

(24) (25)

2.9.2 Epoxy Compounds

Contrary to amines, some structural variations of the diglycidyl ethers of bisphenols, such as the position of the glycidyl groups in the aromatic ring or the presence of either electron-donor or electron-acceptor substituents, have little effect on their reactivity with amines [103–105]. o-Diglycidyl ethers the rate constant of which is higher by a factor of 5 than that of m- and p-isomers are an exception [104]. These data are presented in Tables 7 and 8.

Table 7. The effect of isomerism of diglycidyl ethers of bisphenols on the kinetic parameters of their interaction with 4,4'-diamino-3,3'-dichlorodiphenylmethane [104]

Monomer*	$k_1 \times 10^6$ (l^2 mol^{-2} s^{-1}) T = 400 K	E (kJ mol^{-1})
DGP	18.0	58
DGR	3.0	64
DGH	3.5	64

* DGP, DGR, DGH are diglycidyl ethers of pyrocatechol, resorcinol and hydroquinone, respectively

Table 8. Kinetic parameters for noncatalytic interaction of diepoxy compounds

$$CH_2-CHCH_2O-\bigcirc-X-\bigcirc-OCH_2CH-CH_2$$ with aromatic diamines [105, 106]

X	$k_1 \times 10^7$ (l^2 mol^{-2} s^{-1}) T = 373 K	E (kJ mol^{-1})	$k_1 \times 10^6$ (l^2 mol^{-2} s^{-1}) T = 373 K	E (kJ mol^{-1})
O	7.5	79	6.0	71
S	10.0	64	8.2	59
—	7.4	84	8.4	77
CH$_2$	8.5	82	7.1	68
SO$_2$	11.0	80	15.0	77
CO	6.4	75	8.0	66
N=N	9.9	—	—	—

These data are well described by the Hammet equation with $\varrho = 0.3$. The small value of this parameter for the reaction under study suggests a very weak sensitivity of the reaction site to the type of the bridging groups. To a certain, extent this conclusion could be anticipated, as the reaction site (the epoxy ring) in the glycidyl ethers is separated from the benzene ring and the associated bridge atoms with an oxymethylene group so that the inductive effects is weak [Scheme (26)]:

$$CH_2-CHCH_2O\ (-\bigcirc-\underset{\underset{CH_3}{|}}{\overset{\overset{CH_3}{|}}{C}}-\bigcirc-OCH_2CHCH_2)_n-$$
$$-O-\bigcirc-\underset{\underset{CH_3}{|}}{\overset{\overset{CH_3}{|}}{C}}-\bigcirc-OCH_2CH-CH_2$$
 (26)

The rate of curing of Bisphenol A diglycidyl ether oligomers with amines increases with increasing molecular weight of the oligomer [14, 42, 107]. This fact is quite clear since this increase results in increasing concentration of the hydroxyl groups, which are responsible for a higher reaction rate. Determination of the kinetic parameters in the interaction of these oligomers of varying molecular weight, considering the initial concentration of the hydroxyl groups, indicates that all the oligomers under study are described within the experimental error by the same rate constant.

2.9.3 Reagent Structure and Ineffective Cyclization Reactions

The structure and properties of a network polymer are determined by the relation between the inter- and intramolecular reactions of functional groups. The latter gives rise to ineffective cycles, and has been studied fairly well [3, 5, 70, 71, 81, 85, 108–112]. Their occurrence in the system results in the gelation point being shifted towards higher conversions, higher sol fraction, fewer elastically active chains, and smaller equilibrium modulus of the network.

One quantitative approach to the inclusion of formation of ineffective cycles is to adopt "a network termination probability parameter" within the frameworks of the branching process theory [3, 112]. Another method consists in a computer-assisted modelling of network formation [5, 81, 85, 87, 88, 109].

The relation between the inter- and intramolecular acts is controlled by the reaction conditions such as type and concentration of the reagents, nature of the solvent, temperature, order in which the reagent are mixed, etc. This problem is discussed in much detail elsewhere [3]. In this sub-section, we shall be only concerned with some of the experimental and theoretical studies aimed at revealing a relationship between the reagent structure and ineffective cyclization in the course of curing the epoxy-amine compositions.

An illustrative description of the ineffective cycles is provided by the network polymer random model [81, 87]. The model is underlied by a conception of the network polymer structure as a combination of the accidentally associated cycles with a certain distribution by size. Scheme (27) illustrates the first members of a series of cyclic structures formed by reaction of the diepoxides with diamines (the notation i, j means a ring composed of i diepoxide and j diamine units)

(27)

In this case the 1,1- and 1,0-cycles turned out to be ineffective (defective).

For equilibrium-controlled reactions, the probability of formation of the ineffective cycles depends on their thermodynamic stability. The fraction of such cycles on the stage of polymer formation can be only estimated if we know the relative constant of the rate of formation of the corresponding cycle; however, at present the required experimental data are not available. The qualitative picture can be obtained by use of the molecular Stewart-Brigleb models. Thus, this method has been used to show [81, 113] that diglycidyl ethers of pyrocatechol and Bisphenol A tend to form 1,1-cycles in the interaction with 4,4′-diaminodiphenylsulphone and m-phenylenediamine and the 1,0-cycle becomes very strained; formation of ineffective cycles for the diglycidyl ether of the resorcinol-4,4′-diaminodiphenylsulphone system is improbable due to the high strain in the 1,1-cycle, whereas the 1,0-cycle is impracticable because of geometrical reasons.

A quantitative estimation of the fraction of the 1,1-cycles can be performed through a conformational analysis of the diepoxide molecules, i.e. determination of the number of their conformations in which the end-to-end distance is equal to the spacing between the nitrogen atoms in the corresponding diamine molecules. Such an analysis has given results qualitatively consistent with those obtained in the analysis of the molecular models by the Stewart-Brigleb method [110, 111, 114].

A fairly large number of experiments have been made by means of methods reported elsewhere [2-4, 89, 104, 108, 110, 111, 113, 115-117] to study indirectly the effect of the reagent structure of the epoxy-amine systems on the various physico-mechanical properties of glassy and rubbery polymers and to analyze the ineffective cycles in the model and real systems by use of a variety of techniques. These experiments revealed a relatively high tendency of o-diglycidyl ethers of phenols and dicarbocylic acids to undergo cyclization as compared to the m- and p-isomers. An important factor promoting formation of 1,0-cycles in the diglycidyl ethers is an intramolecular hydrogen bonding in the product of a primary addition of the amine to the diepoxide. It is this factor that seems to be responsible for an enhanced reactivity of o-diglycidyl ethers [118].

2.10 Effect of Solvent Nature

The nonspecific effect of the solvent through the dielectric constant of the reaction medium in the epoxide-amine interaction is very weak [37], which is fully consistent with the molecular mechanism of these reactions. However, even the early studies of these interactions [1, 12, 19, 20] indicate that both the proton- and electron-donor solvents greatly effect the reaction kinetics. The former decrease the induction period and increase the reaction rate; the more pronounced this effect the higher is the solvent acidity. The electron donors decrease both the initial and the postinitial reaction rate; the inhibition effect grows with increasing conversion of epoxy group [13, 14, 45, 55, 56, 119]. Such a behaviour of the solvents is fully consistent with the above-discussed reaction mechanism. The inhibition effect of the electron-donor solvents is due to a decreased concentration of the proton donors in the reaction system as a result of their bonding to the electron-donor molecules of the solvent, initially to amines and then to hydroxyl groups [45].

2.11 Side Reactions

Despite a rather complicated mechanism of the epoxy-amine reaction, the molecular structure of the resulting polymers at moderate temperatures is simple and well described by Scheme (1). This permits a most accurate prediction of the properties of the epoxy-amine polymers both in the glassy and rubbery states [2-6, 88, 120-122].

At the same time, a variety of side reactions take place at elevated (>420 K) temperatures. First of all, it is the reaction of the secondary alcohol groups formed from the epoxy groups studied on model and polyfunctional systems [2-4, 20, 54, 60, 62, 123-125]. It can be concluded that for aromatic amines and epoxides no reaction takes place below 473 K; at higher temperatures one can observe a slow reaction, but the reaction rate is practically the same as for the epoxide-alcohol reaction in the absence of any tertiary amino group, i.e. the formed tertiary aromatic amine has little or no catalytic effect on the reaction. The reaction is catalyzed when primary or secondary aliphatic or aromatic amines containing a nitrogen atom in the heterocycle such as diaminopyridine are used as curing agents. However, even in this case the reactivity of the primary and secondary amino groups is significantly (more than by an order of magnitude) higher than that of the hydroxyl group. It means that the contribution of this reaction at stoichiometric ratios of the functional groups or excess amino groups can be ignored. The role of the reaction becomes quite perceptible in the presence of excess epoxy groups in the starting reaction mixture. Then, this reaction is responsible for the full consumption of the epoxy groups and for the gelation in case of di- or polyepoxy monomers. It is to be noted that the reaction of the hydroxyl and epoxy groups both for the aliphatic and aromatic amines can be detected at a stoichiometric ratio at high reaction temperatures (473 K). This leads to an additional crosslinking of the polymer and, consequently, to an increase in the equilibrium modulus in the rubbery state and to an increase in the glass transition temperature [125]. Very interesting reactions on model systems were reported elsewhere [126].

It has been claimed that the amino alcohol can be formed not only by the usual reaction of an epoxy compound with a secondary amine, but also with a tertiary one [Scheme (28)]

$$PhOCH_2CH\underset{O}{-}CH_2 + (C_2H_5)_3N \rightarrow PhOCH_2\underset{OH}{CH}CH_2\underset{C_2H_5}{\overset{C_2H_5}{N}} + CH_2=CH_2. \quad (28)$$

Amino alcohol can react with epoxy compounds not only at the hydroxyl group as is usually the case, but also by the following reaction [Scheme (29)]

$$PhOCH_2\underset{OH}{CH}CH_2N(C_2H_5)_2 + CH_3OCH_2CH\underset{O}{-}CH_2 \rightleftarrows$$

$$\rightleftarrows PhOCH_2CH\underset{O}{-}CH_2 + CH_3OCH_2\underset{OH}{CH}N(C_2H_5)_2 \quad (29)$$

to yield the new molecules of the amino alcohol and epoxy compound. Equation (29) is in fact a transepoxidation. Its hypothetical mechanism is as follows

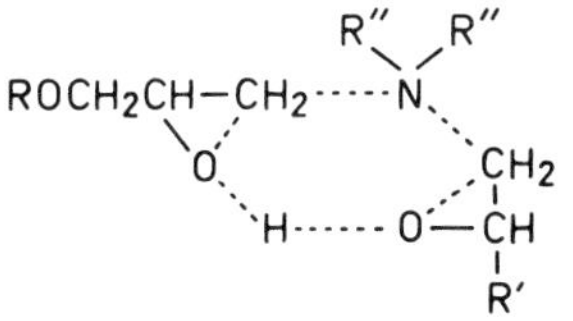

$$\text{ROCH}_2\text{CH}-\text{CH}_2\cdots\overset{R''\diagdown\diagup R''}{N}\cdots\underset{\underset{R'}{CH}}{\overset{CH_2}{|}} \quad (30)$$

This reaction has not yet been found in polyfunctional systems, but the detection is very difficult because the increase in the hydroxyl groups concentration and appearance of unsaturation as a result of reaction (28) is camouflaged by changes taking place in the anionic polymerization of the epoxy groups under the action of tertiary amines (see the following Section). Reaction (29) in a polyfunctional system does not lead to a change of chain length or of site concentration and only redistribution of the sites takes place. Reactions of thermal and thermo-oxidative degradation become important at temperatures above 473 K. A detailed analysis of the degradation products of the polyfunctional and model systems by the mass spectroscopic method [127-130] has shown that the $C^\beta - C^\alpha - N$ bond is the weakest [Scheme (31)]

$$\text{Ph}-O \!\!\mid\!\! CH_2 \!\!\mid\!\! \overset{OH}{\underset{\sim 530K\ \sim 570K \sim 500K}{CH}} \!\!\mid\!\! \overset{H}{CH} - N\!\!<\ \sim 570K$$

The activation energy for dissociation of this bond is 190–230 kJ · mol^{-1}. The corresponding value for the C—O bond is somewhat higher.

The dissociation process is described by a free radical chain mechanism. The thermo-oxidative dissociation is initiated by the oxidation of the aliphatic moieties by a subsequent cleavage of the hydroperoxides formed. With increasing time of oxidation the temperature of the onset of degradation is lower as compared with that for a purely thermal degradation.

In conclusion, a whole class of possible side reactions occur if the epoxy compound contains functional groups capable of reacting with the amines or hydroxyl groups formed during the reaction or directly with the epoxy groups. An example is the aminolysis or alcoholysis of ester bonds during formation of polymers based on diglycidyl ethers of dicarboxylic acids [131-132]. Polymers obtained on the basis of diglycidylurethanes are another example [133-135]. At high temperatures an intermolecular cyclization of the glycidylurethane fragment occurs giving rise to substituted 2-oxazolidone as well as aminolysis of the urethane bond yielding substituted ureas. The reaction rate of aminolysis increases steeply in passing from the aromatic to the aliphatic amine, but even then it is considerably lower compared with the main reaction.

3 Tertiary Amines

Tertiary amines (TA) may be utilized both as independent curing agents for epoxy resins and as catalysts in the reaction of epoxy compounds with alcohols, phenols, carboxylic acids and their anhydrides [1, 12, 14, 19-25, 64, 136-146]. Besides, the three-component mixtures based on TA, epoxy compounds and alcohols have been found to be effective catalysts of such processes as isocyanate trimerization [48] and degradation of various heterochain polymers including sulphur vulcanization agents for rubbers [147, 148]. The mechanism of these various chemical transformations is based on the ability of the epoxy compounds to give high-reactive products with TA. Therefore, a knowledge of the nature of the active sites formed by the reaction of epoxy compounds with TA is the key to understanding the polymerization mechanism of these compounds unter the action of TA.

However, we shall first briefly consider the main kinetic principles of epoxy compound polymerization under the action of TA and the structural peculiarities of the resultant polymers.

3.1 Main Kinetic Principles

The typical form of the kinetic curves is presented in Fig. 9 [14, 149–151]. The curves are distinctly S-shaped. The inductive period becomes shorter and the maximum reaction rate increases upon addition of alcohols. At a given alcohol concentration, the inductive

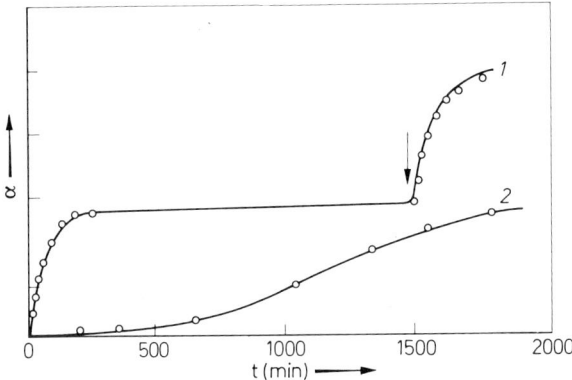

Fig. 9. Kinetic curves of phenylglycidyl ether polymerization in chlorobenzene solution under the action of dimethylbenzylamine and in the presence of isopropyl alcohol at 343 K [14]. 1 $E_0 = A_{3,0} = 1.75$ mol l^{-1}, 2 $H_0 = 0$. The introduction of fresh monomer is shown by an arrow

period disappears and the reaction develops with the maximum rate from the very beginning. The reaction rate increases with increasing alcohol acidity, steric accessibility and basicity of TA [152–153]. The polymerization proceeds up to full conversion of epoxy groups both in the presence and absence of added alcohol. If a fresh portion of monomer is added to the reaction mixture after completion of the reaction, it recommences without any inductive period at the maximum rate [14]. This operation can be repeated many times. With the reaction proceeding below 370 K, no consumption of TA takes place. It follows that generation of the active growing sites is much slower than the chain propagation, and that the proton-donor molecules play an important role in the inhibition. Also, it is obvious that the reaction system has a distinct "living" character, and that the concentration of the active growing sites is only a small fraction of the TA introduced.

The dependence of the reaction rate on the reagent concentration [14, 150–152] is very complicated: variable order with respect to monomer concentration (first order at low and second order at high concentrations); the same order with respect to proton-donor concentration [zero order at low ($<10^{-2}$ mol·l^{-1}) and more than first at high concentrations] and first order with respect to the TA concentration.

All these observations suggest that the polymerization mechanism of the epoxy compounds under the action of TA is very complicated. The same conclusion can be made with respect to the structural peculiarities of the polymer formed.

3.2 Polymer Structure

The structure of the epoxy polymers formed in the curing process has been thoroughly investigated using phenylglycidyl ether polymerization as an example.

It has been established that a polyether mainly with amorphous structure and a small amount (ca. 10%) of a crystalline polymer has been formed [20, 136–139, 154]. The amorphous fraction of the polymer contains approximately one hydroxyl and one unsaturated group per polymer molecule [1, 4, 20, 136, 139, 143, 144, 154–156]. A careful NMR-spectroscopic investigation of the polymers and specially synthesized models of β- and γ-phenoxyallyl alcohols and their methyl ethers has shown that the double bond in the polymer has a vinylidene structure and the primary and secondary hydroxyl end groups are approximately at a ratio of 1:1 [157]. Earlier, other authors [145, 154] have come to an analogous conclusion based on IR-spectroscopy about the nature of the double bonds. The kinetics of accumulation of these groups is the same as the monomer consumption [149, 150] (Fig. 10). Besides, nitrogen traces (ca. 0.2–0.3%)

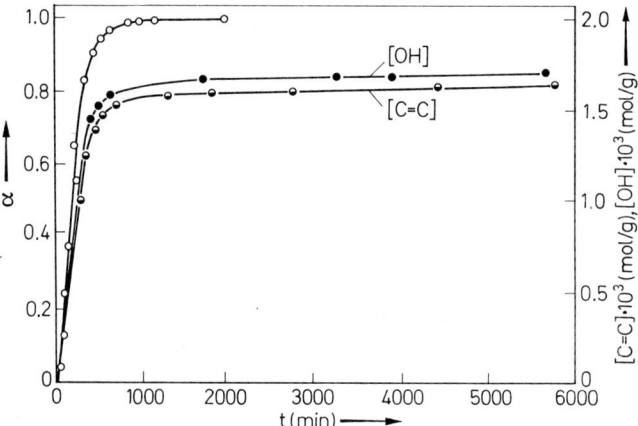

Fig. 10. Kinetic curves of phenylglycidyl ether polymerization (7.0 mol l^{-1}) under the action of dimethylbenzylamine (0.25 mol l^{-1}) in the presence of isopropyl alcohol (0.25 mol l^{-1}) and in the accumulation of hydroxyl and vinyl groups in the system at 353 K [149, 150]

have been detected in the polymer [14, 152, 155]. At high temperatures (420 K), the nitrogen content in the polymer sharply increases [1, 126]. Some peculiarities have been observed in the structure of the polymer obtained without addition of the alcohol [158]. In this case, the method of column adsorption chromatography has been used to separate the polymer into three fractions according to the nature of the end groups: hydroxyl-free- 4.5%; monohydroxyl- 85.4%; dihydroxyl- 10.1%. The polymer obtained in the presence of alcohol additives consists mainly of a monohydroxyl fraction [158].

Molecular-mass characteristics of these polymers are very interesting [1, 14, 20, 136, 154, 155, 158]. An oligomer with a mean degree of polymerization equal to ca. 5 has been formed by polymerization of phenylglycidyl ether. Its molecular-mass characteristics are extremely insensitive either to monomer conversion (Table 9) or to variations of polymerization conditions (Table 10).

Table 9. Dependence of the molecular-mass-distribution of phenylglycidyl ether oligomers on conversion (α) [152, 158]

α	$\bar{M}_W$	$\bar{M}_N$	$\bar{M}_W/\bar{M}_N$
0.10	570	545	1.05
0.36	760	730	1.04
0.58	825	785	1.05
0.76	820	785	1.04
0.98	830	780	1.06
1.00	800	750	1.07

Table 10. Dependence of the molecular-mass distribution of the oligomers on the polymerization conditions of phenylglycidyl ether (PhGE) under the action of dimethylbenzylamine (DMBA) and isopropyl alcohol (IPA) [152, 158]

T (K)	PhGE (mol l^{-1})	DMBA (mol l^{-1})	IPA (mol l^{-1})	$\bar{M}_W$	$\bar{M}_N$	$\bar{M}_W/\bar{M}_N$
298	5.22	1.3	1.3	705	650	1.08
323	5.22	1.3	1.3	680	625	1.09
343	5.22	1.3	1.3	655	595	1.10
343	4.65	0.93	0.09	640	625	1.02
343	7.0	0.23	0.22	735	700	1.05
343	6.82	0.55	—	740	725	1.01
343	5.57	0.11	—	690	685	1.01
343	2.39	4.5	—	590	580	1.02

Reaction conditions: phenylglycidyl ether 5.22 mol · l^{-1}; dimethylbenzylamine 1.3 mol · l^{-1}; isopropanol 1.3 mol · l^{-1}; solvent is chlorobenzene; 343 K.

3.3 Reaction Mechanism

Though the polymerization mechanism of the epoxy compounds under the action of TA has received much attention [1, 12, 19–25, 48, 136–146, 154–156, 159–163], many proposed ideas of the reaction mechanism remained questionable until quite recently. Only the combination of kinetic and structural investigations enables an adequate description of the mechanism of separate elementary stages and the polymerization process as a whole. These investigations have been carried out recently by the author of this review and his coworkers [2, 13, 14, 46, 149–153, 157, 158, 164, 165].

3.3.1 Initiation

First let us consider the initiation process in the presence of proton-donor compounds specially introduced into the system. Two contradictory viewpoints about the reaction mechanism may be distinguished in this case. One of them [138, 139, 145, 154] presupposes a molecular mechanism of the reaction, i.e. a stepwise polyaddition of the epoxy compound to the alcohol group, e.g. according to Scheme (32)

$$R''CH-CH_2 \cdots :OR \atop \underset{O:\cdots}{\diagdown\diagup} \quad \underset{H\cdots NR_3}{|} \quad \rightarrow \quad R''CH-CH_2OR' \atop \underset{OH}{|} \quad + NR_3. \tag{32}$$

The second viewpoint proposed by Eastham and coworkers [161, 162] and then developed by other authors [140–144, 149–153, 155, 157, 158] consists in formation of an active site through a trimolecular transition state [cf. Scheme (33)]

$$R''-CH-CH_2 \cdots :NR_3 \atop \underset{O:\cdots HOR'}{\diagdown\diagup} \quad \rightarrow \quad R'O^-R^3\overset{+}{N}CH_2\underset{OH}{\overset{|}{C}H}-R'' \tag{33}$$

The first viewpoint contradicts the autocatalytic character of the reaction, conductometric measurements in the polymerization system and some other facts (see below). Scheme (33) can be considered as completely experimentally substantiated. The following important proofs were obtained: A direct experimental discovery of a quaternary ammonium alcoholate in the reaction system [142]; a full agreement of the nature of the active propagating site with all the existing kinetic and structural data [14, 149–153, 157, 158]; establishment of the ionic behaviour of the propagating sites by comparison of the kinetic curves of the process with the character of the electric

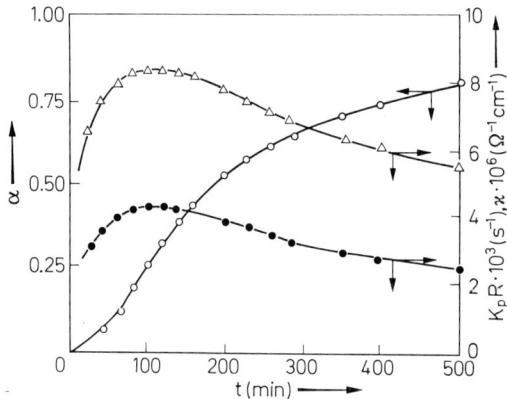

Fig. 11. Kinetic curve of phenylglycidyl ether polymerization in solution of chlorobenzene at 343 K (1) ($E_0 = H_0 = 1.75$ mol l^{-1}, $A_{3,0} = 0.25$ mol l^{-1}) and the change of the relative rate (2) and conductivity (3) of the system during the process [4]. The symbols are the same as in Fig. 9

conductivity change during the reaction (Fig. 11). An identical change of the relative electric conductivity and polymerization rate $(1 - \alpha)(d\alpha/dt) = k_p R^*$, (where K_p is the effective polymerization rate constant, R^* is the concentration of the active propagation sites) unambiguously demonstrates the ionic behaviour of these sites [14, 150–152]. A solution of this kinetic equation has been obtained in combination with the equation for the dissociation constant of the active propagation sites into free ions (K_d) which is a function of the degree of dissociation (β) and equivalent electric conductivity (λ)

$$K_d = \frac{\beta^2 R^*}{1 - \beta}; \quad \beta = \frac{\lambda}{\lambda_\infty}; \quad \lambda = \frac{\varkappa \eta 10^3}{R^*} \tag{34}$$

where $\varkappa$ is the electric conductivity, η is the viscosity of the reaction system at a given time, λ_∞ is the equivalent electric conductivity at infinite dilution. The solution gives an experimental check of the hypothesis about the ionic behaviour of the active propagating particles:

$$\frac{K_p R^*}{\varkappa \eta} = \frac{\varkappa \eta 10^6 k_p^-}{K_d \lambda_\infty^2} + \frac{10^3 k_p^\pm}{\lambda_\infty}, \tag{35}$$

where k_p^- and $k_p^\pm$ are rate constants for the chain propagation the free ion and the ion pair, respectively.

As can be seen from Fig. 12, the experiment is very well described by Eq. (35). Finally, measurements of the electric conductivity of the binary mixtures of the initial reagents have played an important role in substantiating Scheme (35) [149, 152]. The electric conductivity of any binary mixture of the initial reagents (epoxy compounds, TA and alcohol) turned out to be over two orders of magnitude lower than that of the triple system. These experiments show that the interaction of all three reagents is essential for generation of the ionic particles. They have also completely rejected the possibility of formation of active sites directly via interaction of TA with alcohol [139, 156, 163)

$$R_3N + R'OH \rightarrow R'O^- R_3\overset{+}{N}H \cdot \tag{36}$$

Thus, in the presence of alcohols or other proton donors the polymerization of epoxy compounds under the action of TA proceeds according to the anionic mechanism to give quaternary ammonium alcoholate as the active propagating site [Scheme (33)].

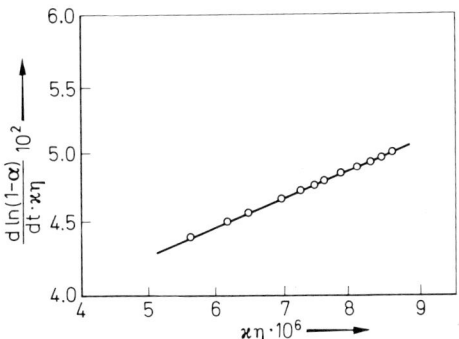

Fig. 12. Kinetic curve of phenylglycidyl ether polymerization (see Fig. 10) according to eq. (35) [14)

It is obvious that the trimolecular transition state is realized through two successive reactions, one of which consists in activation of the epoxy compound during formation of the donor-acceptor complex with alcohol, and the other — in the interaction of this complex with the TA [149-152].

Now the question arises whether the TA can initiate polymerization of epoxy compounds in the absence of any proton-donor compounds. There is no clear solution to this problem. Some investigators [136, 163] believe that the TA itself is capable of opening the epoxy ring to give a zwitter-ion:

$$R'CH\underset{O}{-}CH_2 + R_3N \rightarrow R_3\overset{+}{N}CH_2\underset{R'}{C}HO^-, \qquad (37)$$

which acts as an active growing site. However, this assumption is not correct, since it contradicts the results of conductometric investigations [14, 149-152], for the zwitter-ion is not electrically conductive. This contradiction with Scheme (37) is eliminated to a certain extent by a modification of it consisting in the zwitter-ion transformation according to the reaction of Hofmann intramolecular splitting [163]

$$(CH_3)_3\overset{+}{N}CH_2CH_2O^- \rightarrow CH_2=CHO^-(CH_3)_3\overset{+}{N}H \cdot \qquad (38)$$

The resulting unsaturated alkoxy ion initiates the polymerization process. The reaction of chain transfer to the unstable counterion leads to the appearance of hydroxyl groups in the system; further the initiation occurs in accordance with Eq. (33).

Note that this initiation path through successive transformations by Eqs. (37-39) is in agreement with the existing results of the structural investigations. However, this scheme also should be rejected because it contradicts the existing kinetic data. According to this scheme, the accumulation rate of the hydroxyl groups and double bonds, i.e. the monomer isomerization rate, should be maximum at the very beginning of the reaction. However, this is not the case. As has been already mentioned (see Sect. 3.2), the kinetics of the reaction and of the accumulation of the hydroxyl and

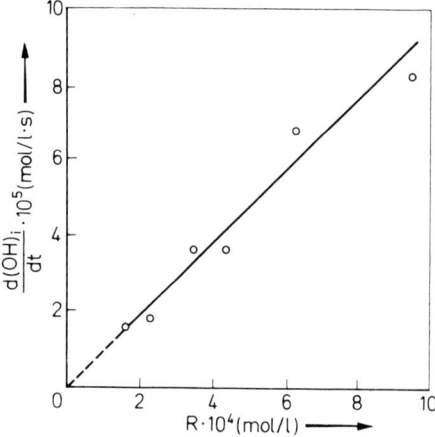

Fig. 13. Dependence of the rate of hydroxyl groups accumulation in the polymerization of phenylglycidyl ether (7 mol l^{-1}) and in the presence of dimethylbenzylamine (0.25 mol l^{-1}) and amyl alcohol (0.25 mol l^{-1}) at 343 K on the concentration of the active growing sites [150]

unsaturated end groups are completely identical (Fig. 10). It means that the rates of all these reactions are governed by the concentration of one and the same active site (Fig. 13). As a matter of fact, the authors of other works [143, 144, 155] adopted a similar way. These authors claim that in the beginning the isomerization of the phenylglycidyl ether (giving phenoxyallyl alcohol) occurs in the system under the action of the TA

$$PhOCH_2CH\underset{O}{-}CH_2 \xrightarrow{R_3N} PhOCH=CHCH_2OH \ . \tag{39}$$

Then it participates in formation of the active growing site by Eq. (33). In reality, however, Eq. (39) does not take place altogether since double bonds of such kind are absent in the polymer [157].

Quite a different approach has been developed in some other works [149, 152]. It has been shown by extrapolation of the dependence of the initial reaction rate on the proton-donor concentration in the system to the zero concentration of the proton-donor that the initial reaction rate is expected to be equal to zero. The authors attempted to carry out precision vacuum cleaning and drying of the reagents and reaction vessels to remove proton-donor impurities. The results of this investigation are given in Fig. 14. As can be seen, even trace amounts of moisture have a great effect on the

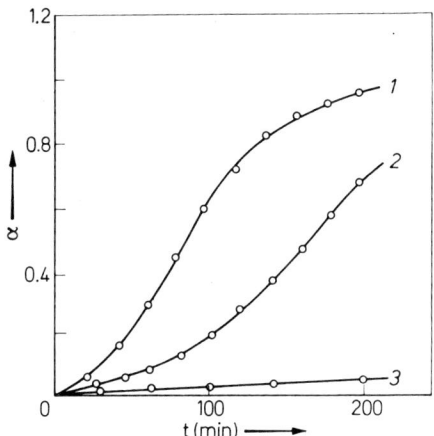

Fig. 14. Kinetic curves of phenylglycidyl ether polymerization (6.67 mol l^{-1}) under the action of dimethylamine (0.66 mol l^{-1}) at 343 K. 1 Pure (chromatography grade) undried phenylglycidyl ether, 2 dried with calcium hydride and dosed in dry argon, 3 precision drying and purification of the reagents in the reaction vessels, reagent dosed in vacuo [149]

reaction rate. The reaction rate falls sharply with a decreasing moisture content in the system. Nevertheless, the authors were unable to stop the reaction completely. This may be impossible in principle because, even in the case of a complete removal of proton-donor compounds from the system and blocking the hydroxyl groups on the surface of the glass vessel by dimethyldichlorosilane treatment, the alkali metal cations on the glass surface may have a catalytic effect.

The effect of the electrophilic assistance to the epoxy ring opening by alkali metal cations is well-known [166, 167]. It is characterized by the quadratic dependence of the epoxy polymerization rate on the alkali metal alcoholate concentration. In this case, a trimolecular transition state is usually realized

$$\begin{array}{c} \overset{\displaystyle OR}{\underset{\displaystyle \vdots}{}} \\ R-CH-CH_2 \\ \diagdown\;\diagup \\ O : \cdots\; Me^+\bar{O}R \end{array} \qquad (40)$$

Thus, the above findings [149] indicate that the first stage of the initiation of the epoxy compound polymerization under the action of the TA even in precision-cleaned systems may take place by the trimolecular mechanism to give quaternary ammonium alcoholates owing to impurity quantities of the proton donor or other electrophilic particles in the system. This reaction will not occur in the absence of such impurities. Upon appearance of the hydroxyl groups in the system due to the chain transfer to the counterion (see Sect. 3.3.3) the reaction according to Scheme (33) with participation of the unsaturated alcohol becomes the main source of the active growing sites.

Thus, the electrophilic impurities act as a trigger mechanism in the aprotic system.

3.3.2 Chain Propagation

As follows from the previous sub-section, polymerization of the epoxy compounds under the action of the TA is a particular case of their anionic polymerization induced by bases; on the stage of chain propagation, this process obeys the well established principles of anionic polymerization of epoxy compounds. This problem has been discussed in a monograph [2] and two excellent reviews [25, 146].

We shall not treat a number of general problems of anionic polymerization such as autosolvation of the ion pairs, cation solvation with the electron-donor chain atoms, the role of the medium etc. Two problems attract our attention: monomer activation during chain propagation and the direction of the epoxy ring opening.

First of all, note the specific features of epoxy compound polymerization under the action of the TA. It consists in accumulation of hydroxyl groups during the reaction; its molar fraction at the end of polymerization may be 20% of the initial concentration of the epoxide groups [158].

This means that the active propagating sites such as free alkoxy ions and ion pairs are solvated with the hydroxyl groups. This must lead to an increase in the reactivity of the solvated ion pair as compared with that of the contact ion pair and decrease in the free ion reactivity, i.e. ultimately to the levelling off of the reactivity differences of these particles.

On the other hand, in the presence of hydroxyl groups the reaction does not involve free epoxy groups, but rather their complexes with hydroxyls, whose reactivity with respect to the nucleophilic reagents is considerably higher [13, 26, 27, 149–152, 166–171]. Many authors consider the epoxy ring activation in such reactions obligatory. During anionic polymerization of epoxy compounds in the absence of proton donors, this role may be played by a cation in the growing ion pair [166–170], the activation effect being the higher, the smaller the cation radius. The results obtained on the model systems are especially interesting (Table 11).

In case of Li^+, the monomer activation effect is so great that an abnormal relation-

Table 11. Constant rates of interaction of ethylene oxide with alkaline salts of carbanions in tetrahydrofurane at 298 K [172)]

Cation	$k_d \times 10^8$ (mol l^{-1})	$k_p^{\pm}$ (l mol s^{-1})	k_p^{-} (l mol^{-1} s^{-1})	$k_p^{-}/k_p^{\pm}$
Li$^+$	17	10^4	200	0.02
Na$^+$	15	13.8	182	13
K$^+$	7.1	6.0	200	33
Rb$^+$	1.1	3.4	195	57
Cs$^+$	1.7	1.7	207	122

ship between the reactivities of the free ion and that of the ion pair is observed, the reactivity of the latter being higher by a factor of 50 compared with the free carbanion. One more fact is of interest: for the biggest Cs$^+$ cation, the $k_p^{-}/k_p^{\pm}$ ratio is as small as 122, which is a few orders of magnitude lower than the usual values for the vinyl monomer polymerization [173)]. It means that even in this case the effect of the epoxy ring activation is still very great.

Abnormal relationships between the reactivities of the free ion and ion pair for the Na$^+$ and K$^+$ counterions were also obtained for the other model systems [174)]. Unfortunately, similar data for tetraalkylammonium cations are absent at present.

The estimation of the reactivities of the free ions and ion pairs directly in the polymerization reaction of phenylglycidyl ether under the action of dimethylbenzylamine in the presence of isopropyl alcohol at 343 K [151)] gave $k_p^{-} = 5.6$ l · mol^{-1} · s^{-1} and $k_p^{\pm} = 0.7$ l · mol^{-1} · s^{-1}. The values of the bimolecular rate constants are given here considering the fact that the activated molecules of the monomer (its complexes with alcohol) take part in the chain propagation reaction.

Table 11 presents one more result important for the chemistry of epoxy compounds, namely: within the experimental error the rate constant of the free ion is the same for all counterions. This means that such strong nucleophilic particles as carbanions (and evidently alkoxy anions) are capable of opening the epoxy ring without additional electrophilic activation. This result explains the apparently contradictory results that, depending on the reaction conditions, either tri- [140–144, 166–171)] or bimolecular kinetics [175–179)] is observed. The bimolecular kinetics also can be explained in terms of the trimolecular mechanism, since proton-donor additives play a dual role.

On one hand, they increase the reaction rate due to an electrophilic assistance for the epoxy ring opening and, on the other, lower the reactivity of the alcoxy anion owing to its solvation and the decrease of its nucleophility. Positive, neutral or even negative effects of the alcohol additives on the reaction rate are governed by the relationship between these two factors. The chain propagation reaction mechanism itself remains trimolecular.

The relationship between the tri- and true bimolecular mechanisms requires additional investigations.

The electrophilic assistance may control not only the kinetic peculiarities of the anionic polymerization of the epoxy compound and the structure of the polymer formed as well.

Two ways of epoxy ring opening are possible for the asymmetric α-oxides:

$$R_jO^- + RCH-CH_2 \underset{O}{\diagdown\diagup} \begin{matrix} \nearrow R_jOCH_2\overset{\overset{R}{|}}{C}HO^- & \text{"normal" or } \beta\text{-opening} \\ \searrow R_jO\underset{\underset{R}{|}}{C}HCH_2O^- & \text{"abnormal" or } \alpha\text{-opening} \end{matrix} \qquad (41)$$

Contrary to the generally accepted point of view [24, 25] by which the "normal" addition products are mainly formed as a result of the basic catalysis, a number of works appeared recently [157, 180–182] reporting on a high yield (up to 1:1) of the "abnormal" addition products. In other words, the epoxy ring opening may occur with equal probability in both directions. For a deeper understanding of these facts, additional quantum-chemical and stereochemical investigations are essential. At present, it may be stated that an increased yield of the "abnormal" addition products is typical of the processes of anionic polymerization of α-oxides in proton media, i.e. if the activated form of the monomer undergoes a nucleophilic attack.

3.3.3 Chain Termination Reactions

A reaction of chain transfer to alcohol or other proton-donor compounds is the most common process among the chain termination reactions in the anionic polymerization of the epoxy compounds in proton media, including the action of TA:

$$R_jO^- + ROH \rightleftarrows R_jOH + RO^- , \qquad (42)$$

where R_j is the polymer residue. As could be expected, the rate and equilibrium of this reaction are governed by the proton-donor acidity. Propargyl and allyl alcohols are active transfer agents, whereas amyl and isopropyl alcohols are ineffective in this respect [149–152].

The second important chain termination reaction characteristic of the catalysis of anionic polymerization of epoxy compounds by the TA consists in the abstraction of the hydrogen atom from the β-carbon atom in the tetraalkylammonium cation by the growing alkoxy anion (β-elimination reaction) [158, 164].

$$R_jO^- \begin{bmatrix} CH_3 \diagdown \diagup CH_3 \\ N-CH_2CHCH_2OPh \\ | \quad | \\ CH_2Ph \quad OH \end{bmatrix}^+ \rightarrow R_jOH + PhCH_2N(CH_3)_2$$

$$+ PhOCH_2\underset{\underset{OH}{|}}{C}=CH_2 \qquad (43)$$

If other alkyl substituents have hydrogen atoms in the β-position, the chain also may be transferred to them. Thus, ethylene and amino alcohols are formed as a result of a similar reaction using triethylamine as TA [see Scheme (28)].

Unsaturated alcohol formed by Eq. (43) has a high acidity, so it can take part both in the chain-transfer reaction by Scheme (42) and in the reinitiation reaction:

$$\text{PhOCH}_2\text{CH}-\text{CH}_2 + \text{PhCH}_2\text{N}(\text{CH}_3)_2 + \text{PhOCH}_2-\underset{\text{OH}}{\text{C}}=\text{CH}_2$$
$$\searrow\!\!\!\!\nearrow$$
$$\text{O}$$

$$\rightarrow \text{CH}_2=\underset{\text{CH}_2\text{OPh}}{\text{C}-\text{O}^-} \quad \left[\underset{\text{CH}_2\text{Ph}\ \ \text{OH}}{\overset{\text{CH}_3\ \text{CH}_3}{\text{N}-\text{CH}_2\text{CHCH}_2\text{OPh}}}\right]^+ \tag{44}$$

Both reactions lead to the appearance of end hydroxyl and unsaturated groups in the polymer. The accumulation rate of these groups is expected to be controlled by the concentration of the active propagating sites (see Fig. 13) in accordance with Scheme (43).

So, from the kinetic point of view Eq. (43) should be considered as a chain transfer to the counterion.

This scheme requires some comments. The vinyl alcohol formed could be expected to form a ketone [164].

$$\underset{\text{CH}_2\text{OPh}}{\text{CH}_2=\text{C}-\text{OH}} \rightleftarrows \underset{\text{CH}_2\text{OPh}}{\text{CH}_3-\text{C}=\text{O}} \tag{45}$$

However, it has been impossible to find a ketone in the reaction medium because the forward rate of Eq. (45) is considerably lower as compared with that of the other procedures of enol consumption. At the same time, a specially synthesized additive of phenoxyacetone leads to a considerable acceleration of phenylglycidyl ether polymerization under the action of TA; the concentration of the keto group during the reaction falls to the full disappearance of this group. The NMR spectroscopic investigation of the products of these reactions has shown that the methyl groups, which are not present in the polymer in the absence of phenoxyacetone additives, appear together with the unsaturated groups of vinylidene structure [164]. This points to the tautomeric transformation of the phenoxyacetone into two enol forms

$$\underset{\text{OH}}{\text{PhOCH}=\text{CCH}_3} \rightleftarrows \overset{\overset{\text{O}}{\|}}{\text{PhOCH}_2-\text{C}-\text{CH}_3} \rightleftarrows \underset{\text{OH}}{\text{PhOCH}_2\text{C}=\text{CH}_2} \tag{46}$$

and confirms the conclusion about faster enol consumption in Eqs. (42) and (44) as compared with the establishment of tautomeric equilibrium.

A small quantity of the polymer without any hydroxy end groups is due to the following reaction [164]

$$\text{R}_j\text{O}^- \left[\underset{\text{CH}_2\text{Ph}\ \ \text{OH}}{\overset{\text{CH}_3\ \text{CH}_3}{\text{N}-\text{CH}_2\text{CH}-\text{CH}_2\text{OPh}}}\right]^+ \rightarrow$$

$$\rightarrow \text{R}_j\text{OCH}_3 + \underset{\text{OH}}{\text{PhCH}_2\overset{\text{CH}_3}{\text{N}}\text{CH}_2\text{CHCH}_2\text{OPh}} \tag{47}$$

The same reaction may be a source of nitrogen traces in the polymer formed by a TA without alkyl substituents having hydrogen atoms in the β-position. If there are such alkyl substituents in the TA, Eq. (28) and the subsequent inhibition by the formed amine alcohols will lead to the appearance of a large amount of nitrogen in the polymer. The use of nitrogen-containing heteroaromatic compounds as TA leads to the same result. The initiation process in these systems is accompanied by opening of the heteroaromatic cycle and building in nitrogen atoms as well as fragments of the polyenic structure into the polymer [152, 164].

The fraction of the dihydroxyl polymer found in the absence of alcohol additives results from the initiation by impurity moisture in the starting reaction system. This problem has been already treated in Sect. 3.3.1. Note, in addition, that the concentration of water in these systems estimated from the dihydroxyl fraction is in a good agreement with the amount obtained from the kinetic data [158].

The features of the molecular-mass characteristics of the polymer such as low molecular mass, a very narrow molecular-mass distribution, and their exceptionally weak dependence on the reaction conditions (temperature and reagent concentration) can be connected only with an increased probability of termination of the growing chain Eq. (43) with increasing chain length. A similar dependence for the reaction with participation of alkoxy anions is known in organic chemistry [183, 184], however, such a behaviour is not quite understood at present.

4 Amine Mixtures

Addition of amines, especially of primary and tertiary ones [1, 13, 22, 185], is often used to modify technological and processing properties of epoxy-amine compositions. Two types of processes occur parallel during polymer formation, i.e. polycondensation and anionic polymerization leading to the formation of aminooxyether and polyether fragments, respectively. A similar situation is also characteristic for the curing of epoxy compounds with aliphatic and nitrogen-containing heteroaromatic primary amines.

It is obvious from the two previous paragraphs that the above-mentioned processes cannot proceed independently. The kinetic features of the processes of curing of the epoxy-amine systems with the mixtures of primary and tertiary amines as well as the mechanism of the simultaneously occurring polymerization and polycondensation [13, 46] will be discussed next.

Kinetic curves of phenylglycidyl ether consumption in the reaction with aniline in the presence of dimethylbenzylamine (DMBA) are presented in Fig. 15a. As can be seen from this figure, addition of DMBA considerably increase the rate of monomer consumption in the initial and especially in the later stages of the reaction. The amount of the monomer consumed during polycondensation and polymerization can be readily determined by a parallel determination of the aniline consumption (Fig. 15b).

Note that the reaction rate in the presence of the mixture of amines (Curve 4) considerably exceeds the rate of monomer consumption if the same amines are reacted separately and are involved in polycondensation (Curve 1) and polymerization (Curve 6); therefore, an obvious synergetic effect is operative. Considering the fact that the initial rate of aniline consumption (Fig. 15b, Curve 4) is practically the

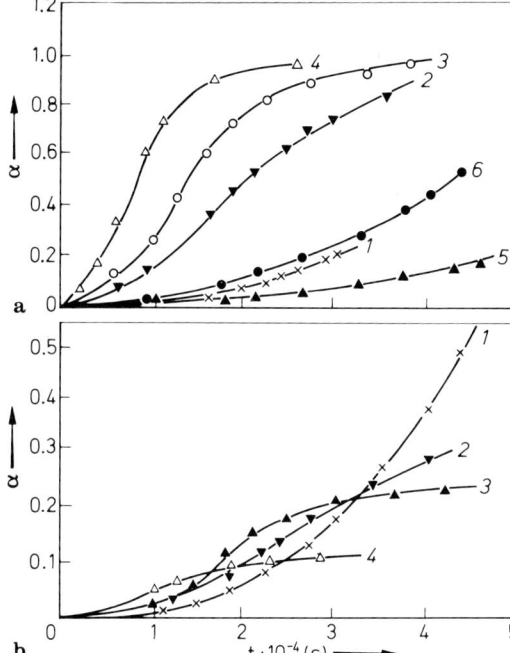

Fig. 15a and b. Kinetic curves of the consumption of phenylglycidyl ether (a) and aniline (b) in the presence of dimethylbenzylamine, mol l^{-1}: 1 0, 2 0.087, 3 0.17, 4 0.35, 5 0.17, 6 0.35, E_0 = 4.38 mol l^{-1}, A_1 = 2.19 mol l^{-1} for curves 1–4, and A_1 = 0 for curves 5 and 6. Solvent — dichlorobenzene, T = 343 K [46].

same and then fairly rapidly falls after a short period of autoacceleration, it becomes clear that the observed reaction acceleration in the presence of amine mixtures is connected with the acceleration of the polymerization reaction.

The acceleration of the initial stage of the epoxy compound polymerization under the action of the TA in the presence of the primary amine is explained by the role of the proton donor of primary proton amine activating the epoxy ring. The initial polymerization rate increases linearly with increasing concentration of the primary amine [46].

At higher conversions, the rate of phenylglycidyl ether and aniline consumption (Curves 1, 4, 6) increases. The polymerization reaction is much more sensitive to the progress of condensation than the condensation process to the polymerization reaction. This kinetic effect becomes clear taking into account the fact that each condensation of epoxide with primary amine gives one hydroxyl group, whereas during polymerization one OH-group is formed per five chain propagations. This suggests that autocatalysis of the polymerization reaction must be more intensive in the presence of the condensation reaction. Note also that the rate constant of polymerization is ca. by one order of magnitude higher than that of condensation [46].

The introduction of TA into the epoxide-primary amine system should lead to the inhibition of polycondensation due to a decrease in the concentration of the free proton-donor molecules due to their bonding to give nonreactive complexes with TA. Therefore, in the general case the curing rate in the presence of amine mixtures can be both higher (as in the above case) and lower as compared with the primary amine only [1, 22, 46]. The final result depends on the relationship between the effective rate constants of polycondensation and polymerization reactions, nature of amines and their concentration.

TA can be effetive additives for controlling the reaction rate in the curing of epoxy compounds with primary amines.

5 Conclusion

Although the present review is the most detailed one as compared with earlier publications, a number of relevant problems have only been touched upon in passing, and some of them have been entirely omitted, as they are dealt with in other chapters (high-temperature transformation of epoxy compounds in the presence of TA [186] and network structure [187]) of this volume.

Analysis of the existing experimental data on kinetics, thermodynamics and mechanism of the epoxy compound curing with amines shows that the available experimental data may be consistently interpreted as a whole only in terms of the trimolecular mechanism of the reactions of epoxy ring opening under the action of the moderately strong nucleophilic reagents, assuming a preliminary ring activation with some electrophilic reagents. Different proton-donor compounds, alkali metal cations or ammonium ions can act as such reagents. In this case, even such weak acids as primary and secondary amines can act as proton-donors. The trimolecular mechanism is also operative in the curing of epoxy compounds under the action of TA even in systems comparatively free of proton donors both on the initiation and chain propagation stages, because the hydroxyl groups are formed during the reaction.

The second important feature of the reaction mechanism of the epoxy compound curing under the action of amines (primary, secondary and tertiary) and their mixtures consists in formation of various hetero-, auto-, inter- and intramolecular donor-acceptor complexes between the components of the reaction system — the starting substances and reaction products. Consideration of this complex formation can adequately explain the reaction kinetics.

These ideas also have been very useful in understanding the interaction of the epoxy compounds with other nucleophilic reagents such as phenols [188] and carboxylic acids [189].

In the present review, a first attempt has been made to discuss in more detail the kinetic features of the deep conversion stage of curing of epoxy-amine systems. However, this problem is not restricted to a given system and is of a general significance to the processes of formation of all network polymers. Some aspects of this problem have been considered earlier [3, 5, 64, 70, 71, 76–85]. Its special significance, as has been shown in recent years [3, 76–78, 96, 97], is due to the fact that the kinetic features of the deep conversion stages of the curing processes can considerably govern the structure and properties of the networks formed in the glassy state. The structure and properties of the polymers in the rubbery state [3, 64] are less sensitive to the kinetic features. Unfortunately, we have to note some gap between the experimental investigations of the various aspects of this interesting and important kinetic and structural problem of the network polymer formation and the available theoretical predictions.

As can be seen from the present review, in the last two decades an extensive experimental and theoretical development of our understanding of the formation of epoxy-amine systems could be recorded. Many other aspects of the problem under discussion may be considered as completely solved or close to solution. At the same time, many other problems are still awaiting their solution.

6 References

1. Lee, C., Neville, K.: Handbook of Epoxy Resins, Translation into Russian, Moskva: Energiya 1973
2. Rozenberg, B. A., Oleinik, E. F.: Usp. Khimii 53, 273 (1984)
3. Irzhak, V. I., Rozenberg, B. A., Enikolopyan, N. S.: Setchatye polimery: sintez, struktura, svoistva, Moskva: Nauka, 1979
4. Rozenberg, B. A.: Dokl. I Vsesoyuznoi konf. po khimii i fizikokhimii polimerizatsionno-sposobnykh oligomerov. Chernogolovka: Izd. IKhF AN SSSR p. 392 (1977)
5. Enikolopyan, N. S.: Dokl. I Vsesoyuzn. konf. po khimii i fizikokhimii polimerizatsionno-sposobnykh oligomerov. Chernogolovka: Izd. IKhF AN SSSR p. 87 (1977)
6. Rozenberg, B. A., Oleinik, E. F., Irzhak, V. I.: Zh. Vsesoyuz. Khim. Obch. im. D. I. Mendeleeva 23, 272 (1978)
7. Kardashov, D. A.: Konstruktsionnye klei. Moskva: Khimya, 1980
8. Kardashov, D. A.: Epoksidnye klei. Moskva: Khimiya, 1973
9. Shields, J.: Adhesive Materials, Russian Translation, ed. V. P. Batizat, Moscow: Mashinostroenie 1980
10. Kardashov, D. A., Petrova, A. P.: Polimernye klei. Moskva: Khimiya, 1983
11. Blagonravova, A. A., Nepomnyashchii, A. I.: Lakovye epoksidnye smoly, Moskva: Khimiya, 1980
12. Paquen, A. M.: Epoxydverbindungen und Epoxydharze, Berlin, Springer, 1958
13. Rozenberg, B. A., Enikolopyan, N. S.: Polimery, 25, 215 (1980)
14. Rozenberg, B. A.: in Kompositsionnye polimernye materialy. Kiev: Naukova dumka 1975 p. 39
15. Rozenberg, B. A., Irzhak, V. I.: in Structura i svoistva polimernykh materialov. Riga: Zinatne, 1979, p. 12
16. Rozenberg, B. A., Enikolopyan, N. S.: Zh. Vsesoyuz. Khim. Obch. im. D. I. Mendeleeva 23, 298 (1978)
17. Rozenberg, B. A.: Fibre Sci. Technol. 19, 77 (1983)
18. Pershin, V. A. et al.: Dokl. Akad. Nauk SSSR 256, 421 (1981)
19. Malinovskii, M. S.: Okisi olefinov i ikh proizvodnye. Moskva: Goshimizdat, 1961
20. May, C. A., Tanaka, Y.: Epoxy Resins. Chemistry and Technology. New York: Decker, 1973
21. Furukawa, J., Saegusa, T.: Polymerization of Aldehydes and Oxides, New York: Wiley, 1963
22. Partansky, A. M.: Epoxy Resins, Washington (1970)
23. Brojer, Z., Hertz, Z., Penchek, P.: Zywice epoksydowe. Warszawa: WNT, 1972
24. Parker, R. E., Isaacs, N. S.: Chem. Rev. 59, 737 (1959)
25. Entelis, S. G., Kasanskii, K. S.: Uspekhi khimii i fiziki polimerov. Moskva, Khimiya, 1970 p. 324
26. Smith, I. T.: Polymer 2, 95 (1961)
27. Gough, L. J., Smith, I. T.: J. Appl. Polym. Sci. 5, 86 (1961)
28. Arutyunyan, Kh. A. et al.: Dokl. Akad Nauk SSSR 214, 832 (1974)
29. Arutyunyan, Kh. A. et al.: Vysokomol. soedin. A 17, 1647 (1977)
30. Arutyunyan, Kh. A. et al.: Zh. Fiz. Khim. 48, 2896 (1974)
31. Mutin, I. I. et al.: Izv. Akad. Nauk SSSR, ser. Khim., p. 2828 (1977)
32. Becker, H.: Einführung in die Elektrontheorie organisch-chemischer Reaktionen, Berlin: Springer, 1964
33. Vladimirov, L. V. et al.: Vysokomol. soedin. A 22, 225 (1980)
34. Arutyunyan, Kh. A. et al., in: "II Vsesoyuznaya konf. po epoksidnym monomeram i epoksidnym smolam". Dnepropetrovsk, p. 129 (1974)

35. Arutyunyan, Kh. A. et al.: Zh. Fiz. Khim. *50*, 2016 (1976)
36. Vladimirov, L. V., Zelenetskii, A. N., Oleinik, E. F.: Vysokomol. soedin. *A 19*, 2104 (1977)
37. Vedenyapina, N. S. et al.: Izv. Akad. Nauk SSSR, ser. Khim., 1956 (1976)
38. Pimentel, J., Mc-Klellan, O.: Vodorodnaya svyaz, Moskva: Mir 1964, p. 303
39. Lady, H., Whetsel, K.: J. Phys. Chem. *71*, 1421 (1967)
40. Tiger, R. P.: in Mechanizmy geteroliticheskikh reaktsii. Moskva: Nauka, 1976, p. 177
41. Enikolopov, N. S.: "Composite Materials — II", 1977, Reports of the First Soviet-Japanese Symposium on Composite Materials. Moscow: University Press, 1979, p. 42
42. Arutyunyan, Kh. A. et al.: Vysokomol. soedin. *A 17*, 289 (1975)
43. Noskov, A. M.: Zh. prikl. spektroskopii *22*, 246 (1975)
44. Arutyunyan, Kh. A. et al.: Dokl. Akad. Nauk SSSR *212*, 1128 (1973)
45. Mutin, I. I. et al.: Kinetika i kataliz *20*, 1567 (1979)
46. Mutin, I. I. et al.: Vysokomol. soedin. *A 22*, 1828 (1980)
47. Tovmasyan, M. A.: PhD Thesis. Moskva (1982)
48. Tiger, R. P.: DSc Thesis. Moskva (1979)
49. Kuznetsova, V. P. et al.: Dokl. Akad. Nauk SSSR *225*, 605 (1976)
50. Horie, K. et al.: J. Polym. Sci. A 1, *8*, 1357 (1970)
51. Arutyunyan, Kh. A.: PhD Thesis. Chernogolovka (1974)
52. Klute, C. H., Viehmann, W.: J. Appl. Polym. Sci. *5*, 186 (1961)
53. Kuznetsova, V. P. et al.: Dokl. Akad Nauk SSSR *226*, 1109 (1976)
54. Charlesworth, J.: J. Polym. Sci., Polym. Chem. Ed. *18*, 621 (1980)
55. Fedtke, M., Tarnow, M.: Plaste Kautsch. *8*, 444 (1981)
56. Fiala, V., Lidařík, M.: Makromol. Chem. *154*, 81 (1972)
57. Dobáš, I., Eichler, J.: Collect. Czech. Chem. Commun. *38*, 2602 (1973)
58. Eichler, J., Dobáš, I.: Collect. Czech. Chem. Commun. *38*, 3461 (1973)
59. Dobáš, I., Eichler, J.: Collect. Czech. Chem. Commun. *38*, 3279 (1973)
60. Dušek, K., Bleha, M., Luňák, S.: J. Polym. Sci., Polym. Chem. Ed. *15*, 2393 (1977)
61. Luňák, S., Dušek, K.: J. Polym. Sci., Polym. Symp. *53*, 45 (1975)
62. Bokare, U. M., Gandhi, K. S.: J. Polym. Sci., Polym. Chem. Ed. *18*, 857 (1980)
63. Kogarko, N. S. et al.: Vysokomol. soedin. *A 20*, 756 (1978)
64. Dušek, K., in: "Rubber-Modified Thermosets", C. K. Riew and J. K. Gillham, editors, Adv. Chem. Ser. 208, p. 000, Am. Chem. Soc. 1984
65. Dušek, K., Ilavský. M., Luňák, S.: Polym. Sci., Polym. Symp. *53*, 29 (1975)
66. Burchard, W., Bantle, S., Zahir, S. A.: Makromol. Chem. *182*, 143 (1981)
67. Berni, R. J., Benerito, R. R., Ziifle, H. M.: J. Phys. Chem. *69*, 1882 (1965)
68. Chang, T. D., Carr, S. H., Brittain, J. O.: Polym. Eng. Sci. *22*, 1213 (1982)
69. Arutyunyan, Kh. A. et al.: Vysokomol. soedin. *A 16*, 2115 (1974)
70. Rozenberg, B. A.: in Kinetika i mekhanism makromolekulyarnykh reaktsyi. Chernogolovka: IKhF AN SSSR, 1983, p. 39
71. Irzhak, V. I., Rozenberg, B. A.: Vysokomol. soedin., *A 27*, 1795 (1985)
72. Gordon, M., Ward, T. S., Whitney, R. F.: in Polymer Networks: Structure and Mechanical Properties ed. Chompff A. U., Newmann, S. New York: Plenum Press, 1971, p. 1
73. Enns, J. B., Gillham, J. K., Small, R.: Polym. Preprints *22*, 123 (1981)
74. Luňák, S., Vladyka, J., Dušek, K.: Polymer *19*, 931 (1978)
75. Gupta, A. et al.: J. Appl. Polym. Sci. *28*, 1011 (1983)
76. Salamatina, O. B. et al.: Vysokomol. soedin. *A 23*, 2360 (1981)
77. Oleinik, E. F. et al.: Chim. fizika *3*, (1984)
78. Oleinik, E. F.: Pure Appl. Chem. *53*, 1567 (1981)
79. Pakhomova, L. K. et al.: Vysokolol. soedin. *B 20*, 554 (1978)
80. Berlin, Al. Al., Oshmyan, V. G.: Vysokomol. soedin. *A 18*, 2282 (1976)
81. Topolkaraev, V. A. et al.: Vysokomol. soedin. *A 21*, 1515 (1979)
82. De Gennes, P.: Idei skeilinga v fizike polimerov Moskva, Mir, 368 (1982)
83. De Gennes, P. G.: J. Chem. Phys. *76*, 3316 (1982)
84. De Gennes, P. G.: J. Chem. Phys. *76*, 3322 (1982)
85. El'yashevich, A. M., Saakyan, L. L.: V sb.: Kinetika i mekhanizm makromolekulyarnykh reaktsyi. Chernogolovka: IKhF AN SSSR, 1983, p. 37

86. Sunagawa, S., Doi, M.: Polymer J. 7, 604 (1975)
87. Topolkaraev, V. A. et al.: Dokl. Akad. Nauk SSSR 225, 1124 (1975)
88. Topolkaraev, V. A. et al.: Mekhanika kompositnikh materialov, 195 (1981)
89. Bogdanova, L. M. et al.: Vysokomol. soedin. A 18, 1100 (1976)
90. Bogdanova, L. M. et al.: Vysokomol. soedin. B 21, 683 (1979)
91. Bogdanova, L. M. et al.: Polym. Bull. 4, 119 (1981)
92. Dušek, K. et al.: Polymer 19, 393 (1978)
93. Dušek, K., Ilavský, M.: in "Elastomers and Rubber Elasticity", J. E. Mark and J. Lal (eds.) ACS, Symp. Ser. Vol 193, Am. Chem. Soc.: Washington 1982, p. 403
94. Dušek, K., Ilavský, M.: J. Polym. Sci., Polym. Phys. Ed. 21, 1323 (1983)
95. Ilavský, M., Bogdanova, L. M., Dušek, K.: J. Polym. Sci., Polym. Phys. Ed. (1983)
96. Irzhak, V. I. et al.: Dokl. AN SSSR 263, 630 (1982)
97. Berlin, A. A. et al.: Akrilovye oligomery i materialy na ikh osnove. Moskva, Khimiya, (1983)
98. Arutyunyan, Kh. A. et al.: Dokl. Akad. Nauk SSSR 222, 657 (1975)
99. Davtyan, S. P. et al.: Vysokomol. soedin. A 19, 2726 (1977)
100. Davtyan, S. P. et al.: in Vtoraya Vsesoyuznaya konf. po epoksidnym monomeram i epoksidnym smolam. Dnepropetrovsk, p. 172 (1974)
101. Surkov, N. F. et al.: Dokl. Akad. Nauk SSSR 228, 141 (1976)
102. Scribner, J. P., Miller, J. R.: J. Org. Chem. 32, 2348 (1967)
103. Komarov, B. A. et al.: Plast. massy No 6, 44 (1984)
104. Efremova, A. I. et al.: Izv. Akad. Nauk SSSR, ser. khim., p. 1112 (1979)
105. Redkina, N. K., Dzhavadyan, E. A., Rozenberg, B. A.: Vysokomol. soedin. A 21, 780 (1979)
106. Dzhavadyan, E. A. et al.: Polym. Bull. 4, 479 (1981)
107. Blyakhman, E. M. et al.: Vysokomol. Soedin. A 16, 1031 (1974)
108. Topolkaraev, V. A. et al.: Vysokomol. soedin. A 21, 1655 (1979)
109. Topolkaraev, V. A. et al.: Dokl. Akad. Nauk SSSR 226, 880 (1976)
110. Chepel, L. M. et al.: Dokl. Akad. Nauk SSSR 266, 415 (1982)
111. Chepel, L. M. et al.: Vysokomol. soedin. A 24, 1646 (1982)
112. Salamatina, O. B. et al.: Vysokomol. soedin. A 25, 179 (1983)
113. Nata, N., Kumanotani, J.: J. Appl. Polym. Sci. 15, 2371 (1971)
114. Topolkaraev, B. A. et al.: Vysokomol. soedin. A 22, 1013 (1980)
115. Ore, S., Tijugum, O. G.: Acta Chem. Scand. 24, 2397 (1970)
116. Yurechko, N. A. et al.: Vysokomol. soedin. A 19, 357 (1977)
117. Berlin, Al. Al. et al., in: Mendeleevskii s'ezd po obshchei i prikladnoi khimii. Moskva, Nauka (1974)
118. Salamatina, O. B., Tarasova, G. M., Ivanov, V. V.: Izv. Akad. Nauk SSSR, ser. khim. p. 1289 (1978)
119. Blyakhman, E. M., Shevchenko, Z. A., Alekseeva, E. M.: Vysokomol. soedin. A 18, 2208 (1976)
120. Ponomareva, T. I., Irzhak, V. I., Rozenberg, B. A.: Vysokomol. soedin. A 20, 597 (1978)
121. Askadskii, A. A. et al.: Vysokomol. soedin. A 25, 56 (1983)
122. Salamatina, O. B. et al.: Vysokomol. soedin. A 25, 179 (1983)
123. Kwei, T. K.: J. Polym. Sci. A 1, vol. 1, 2971 (1963)
124. Anderson, H. C.: SPE J. 16, 1241 (1960)
125. Zhorina, L. A. et al.: Vysokomol. soedin. B 21, 811 (1979)
126. Fedtke, M., Tarnow, M.: Plaste Kautschuk 30, 70 (1983)
127. Zarkhin, L. S. et al.: Dokl. Akad. Nauk SSSR 239, 360 (1978)
128. Zhorina, L. A. et al.: Vysokomol. soedin. A 23, 2799 (1981)
129. Zarkhina, T. S. et al.: Vysokomol. soedin. A 24, 584 (1982)
130. Zarkhina, T. S. et al.: Vysokomol. soedin. A 24, 2479 (1982)
131. Chepel, L. M. et al.: Vysokomol. soedin. A 25, 410 (1983)
132. Chepel, L. M. et al.: Vysokomol. soedin. A 26, 362 (1984)
133. Sorokin, M. F., Shode, L. G., Onosova, L. A.: Lakokr. materialy No 6, 6 (1978)
134. Pankratov, V. A., Frenkel, Ts. M., Fainleib, A. M.: Uspekhi khimii 52, 1018 (1983)
135. Sorokin, M. F.: Izv. vuzov, ser. Khimiya i khim. tekhnologiya No 8, 1220 (1978)
136. Narracott, E. S.: Brit. Plastics 26, 120 (1953)
137. Shechter, L., Wynstra, J.: Ind. Eng. Chem. 48, 86 (1956)
138. Lidařík, M., Starý, S., Mleziva, J.: Vysokomol. soedin. 5, 1748 (1963)

139. Tanaka, Y., Tomio, N., Kakiuchi, H.: J. Macromol. Sci. A 1, 471 (1967)
140. Sorokin, M. F., Shode,L. G.: Zh. org. khim. 2, 1463 (1966)
141. Sorokin, M. F., Shode, L. G.: Zh. org. khim. 2, 1469 (1966)
142. Sorokin, M. F., Shode, L. G., Shteinpress, A. B., Finyakina, L. N.: Kinetika kataliz 9, 548 (1968)
143. Sorokin, M. F., Shode, L. G., Shteinpress, A. B.: Vysokomol. soedin. A 13, 747 (1971)
144. Sorokin, M. F., Shode, L. G., Shteinpress, A. B.: Vysokomol. soedin. A 14, 309 (1972)
145. Tanaka, Y., Kakiuchi, H.: J. Macromol. Chem. 1, 307 (1966)
146. Kazanskii, K. S., in: Khimiya i tekhnologiya vysokomolekulyarnykh soedinenii 9, 5 (1977)
147. Antipova, V. F., Melamed, V. I.: USSR Pat. No 2500450 (1969)
148. Andreev, V. N., Korotysheva, V. V., Rappoport, L. Ya., Petrov, G. N.: Kauchuk rezina, No 3, 11 (1982)
149. Kushch, P. P., Komarov, B. A., Rozenberg, B. A.: Vysokomol. soedin. A 21, 1697 (1979)
150. Komarov, B. A., Kushch, P. P., Rozenberg, B. A.: Vysokomol. soedin. A 26, 1732 (1984)
151. Komarov, B. A., Kushch, P. P., Rozenberg, B. A.: Vysokomol. soedin. A 26, 1747 (1984)
152. Kushch, P. P.: PhD Thesis, Moscow (1981)
153. Kushch, P. P., Komarov, B. A.: V sb.: Vsesoyuznaya konf. po khimii i fiziko-khimii oligomerov. Alma-Ata, p. 204 (1979)
154. Tanaka, Y., Kakiuchi, H.: J. Polym. Sci. A 1, 109 (1966)
155. Sorokin, M. F., Shode, L. G., Shteinpress, A. B.: Vysokomol. soedin. B 11, 172 (1969)
156. Pirozhnaya, L. N.: Vysokomol. soedin. A 12, 2446 (1970)
157. Kushch, P. P. et al.: Vysokomol. soedin. B 21, 708 (1979)
158. Kushch, P. P. et al.: Vysokomol. soedin. A 22, 2012 (1980)
159. Bruin, P.: Kunststoff-Rundschau 6, 485 (1959)
160. Sdyakin, V. N.: Vysokomol. soedin A 14, 979 (1972)
161. Eastham, A. M., Darwent, B. de B., Beaubien, P. E.: Can. J. Chem. 29, 575 (1951)
162. Eastham, A. M., Darwent, B. de B.: Can J. Chem. 29, 585 (1951)
163. Kazanskii, K. S., Solov'yanov, A. A., Entelis, S. G.: "Advances in Ionic Polymerization". Warsaw, p. 77 (1975)
164. Kushch, P. P., Komarov, B. A., Rozenberg, B. A.: Vysokomol. soedin. A 24, 312 (1982)
165. Rozenberg, B. A., Kushch, P. P., Komarov, B. A.: V sb.: Uspekhi v oblasti ionnoi polimerizatsii. Ufa, p. 45 (1979)
166. Lebedev, N. N., Baranov, Yu. I.: Vysokomol. soedin. 8, 198 (1966)
167. Lebedev, N. N. et al.: Teor. i eksperim. Khimiya 4, 203 (1968)
168. Patat, F., Wojteck, B.: Makromol. Chem. 37, 2 (1960)
169. Patat, F., Erlmeyer, R.: Makromol. Chem. 91, 231 (1966)
170. Itakura, I., Patat, F.: Makromol. Chem. 68, 158 (1963)
171. Tanaka, Y.: J. Macromol. Sci. vol. A 1, 471 (1968)
172. Solovyanov, A. A., Kazanskii, K. S.: Vysokomol. soedin. A 16, 595 (1974)
173. Szwarz, M.: Carbanions, Living Polymers and Electron Transfer Processes, New York: Wiley, 1968
174. Uidal, B. et al.: IUPAC 1st Intern. Symp. on Polymerization of Heterocycles. Warsaw-Yablonna, p. 87 (1975)
175. Gee, G. et al.: J. Chem. Soc. 1338 (1956)
176. Gee, G., Higginson, W. C. E., Marral, G. T.: J. Chem. Soc., 1345 (1959)
177. Gee, G. et al.: J. Chem. Soc. 4298 (1961)
177. Gee, G. et al.: J. Chem. Soc. 4298 (1961)
178. Steiner, E., Belletier, R. R., Trucks, R. O.: J. Amer. Chem. Soc. 86, 4678 (1964)
179. Sorokin, M. F., Shode, L. G.: Vysokomol. soedin. 1, 1487 (1959)
180. Ishimori, M. et al.: Makromol. Chem. 115, 103 (1968)
181. Laird, R. M., Parker, R. E.: J. Am. Chem. Soc. 83, 4277 (1961)
182. Komarov, B. A., Volkov, V. P., Boiko, G. N., Naidovskii, E. S., Rozenberg, B. A.: Vysokomol. soedin. A 25, 1431 (1983)
183. Ugelstad, I., Berge, A., Ziston, H.: Acta Chem. Scand. 19, 208 (1965)
184. Thomassen, L. M., Elingsen, T., Ugelstad, I.: Acta Chem. Scand. 25, 3024 (1971)

185. Smirnov, Yu. N. et al., in: Modifikatsiya, struktura i svoistva epoksidnykh polimerov. Kazan: Kazan. Inzhenero-Stroitel. Inst. 1976, p. 30
186. Fedtke, M.: Adv. Polym. Sci. in press
187. Dušek, K.: Adv. in Polym. Sci. *78/79* (1986)
188. Volkov, V. P. et al.: Vysokomol. soedin. *A 24*, 2520 (1982)
189. Saratovskikh, E. A. et al.: Vysokomol. soedin. *B 24*, No 5, 365 (1982)

Editor: K. Dušek
Received May 30. 1985

Author Index Volumes 1–75

Allegra, G. and *Bassi*, I. W.: Isomorphism in Synthetic Macromolecular Systems. Vol. 6, pp. 549–574.
Andrews, E. H.: Molecular Fracture in Polymers. Vol. 27, pp. 1–66.
Anufrieva, E. V. and *Gotlib*, Yu. Ya.: Investigation of Polymers in Solution by Polarized Luminescence. Vol. 40, pp. 1–68.
Apicella, A. and *Nicolais*, L.: Effect of Water on the Properties of Epoxy Matrix and Composite. Vol. 72, pp. 69–78.
Apicella, A., *Nicolais*, L. and *de Cataldis*, C.: Characterization of the Morphological Fine Structure of Commercial Thermosetting Resins Through Hygrothermal Experiments. Vol. 66, pp. 189–208.
Argon, A. S., *Cohen*, R. E., *Gebizlioglu*, O. S. and *Schwier*, C.: Crazing in Block Copolymers and Blends. Vol. 52/53, pp. 275–334
Arridge, R. C. and *Barham*, P. J.: Polymer Elasticity. Discrete and Continuum Models. Vol. 46, pp. 67–117.
Aseeva, R. M., *Zaikov*, G. E.: Flammability of Polymeric Materials. Vol. 70, pp. 171–230.
Ayrey, G.: The Use of Isotopes in Polymer Analysis. Vol. 6, pp. 128–148.

Bässler, H.: Photopolymerization of Diacetylenes. Vol. 63, pp. 1–48.
Baldwin, R. L.: Sedimentation of High Polymers. Vol. 1, pp. 451–511.
Balta-Calleja, F. J.: Microhardness Relating to Crystalline Polymers. Vol. 66, pp. 117–148.
Barton, J. M.: The Application of Differential Scanning Calorimetry (DSC) to the Study of Epoxy Resins Curing Reactions. Vol. 72, pp. 111–154.
Basedow, A. M. and *Ebert*, K.: Ultrasonic Degradation of Polymers in Solution. Vol. 22, pp. 83–148.
Batz, H.-G.: Polymeric Drugs. Vol. 23, pp. 25–53.
Bell, J. P. see *Schmidt*, R. G.: Vol. 75, pp. 33–72.
Bekturov, E. A. and *Bimendina*, L. A.: Interpolymer Complexes. Vol. 41, pp. 99–147.
Bergsma, F. and *Kruissink*, Ch. A.: Ion-Exchange Membranes. Vol. 2, pp. 307–362.
Berlin, Al. Al., *Volfson*, S. A., and *Enikolopian*, N. S.: Kinetics of Polymerization Processes. Vol. 38, pp. 89–140.
Berry, G. C. and *Fox*, T. G.: The Viscosity of Polymers and Their Concentrated Solutions. Vol. 5, pp. 261–357.
Bevington, J. C.: Isotopic Methods in Polymer Chemistry. Vol. 2, pp. 1–17.
Bhuiyan, A. L.: Some Problems Encountered with Degradation Mechanisms of Addition Polymers. Vol. 47, pp. 1–65.
Bird, R. B., *Warner*, Jr., H. R., and *Evans*, D. C.: Kinetik Theory and Rheology of Dumbbell Suspensions with Brownian Motion. Vol. 8, pp. 1–90.
Biswas, M. and *Maity*, C.: Molecular Sieves as Polymerization Catalysts. Vol. 31, pp. 47–88.
Biswas, M., *Packirisamy*, S.: Synthetic Ion-Exchange Resins. Vol. 70, pp. 71–118.
Block, H.: The Nature and Application of Electrical Phenomena in Polymers. Vol. 33, pp. 93–167.
Bodor, G.: X-ray Line Shape Analysis. A. Means for the Characterization of Crystalline Polymers. Vol. 67, pp. 165–194.
Böhm, L. L., *Chmeliř*, M., *Löhr*, G., *Schmitt*, B. J. and *Schulz*, G. V.: Zustände und Reaktionen des Carbanions bei der anionischen Polymerisation des Styrols. Vol. 9, pp. 1–45.

Bovey, F. A. and *Tiers, G. V. D.:* The High Resolution Nuclear Magnetic Resonance Spectroscopy of Polymers. Vol. 3, pp. 139–195.
Braun, J.-M. and *Guillet, J. E.:* Study of Polymers by Inverse Gas Chromatography. Vol. 21, pp. 107–145.
Breitenbach, J. W., Olaj, O. F. und *Sommer, F.:* Polymerisationsanregung durch Elektrolyse. Vol. 9, pp. 47–227.
Bresler, S. E. and *Kazbekov, E. N.:* Macroradical Reactivity Studied by Electron Spin Resonance. Vol. 3, pp. 688–711.
Bucknall, C. B.: Fracture and Failure of Multiphase Polymers and Polymer Composites. Vol. 27, pp. 121–148.
Burchard, W.: Static and Dynamic Light Scattering from Branched Polymers and Biopolymers. Vol. 48, pp. 1–124.
Bywater, S.: Polymerization Initiated by Lithium and Its Compounds. Vol. 4, pp. 66–110.
Bywater, S.: Preparation and Properties of Star-branched Polymers. Vol. 30, pp. 89–116.

Candau, S., Bastide, J. and *Delsanti, M.:* Structural, Elastic and Dynamic Properties of Swollen Polymer Networks. Vol. 44, pp. 27–72.
Carrick, W. L.: The Mechanism of Olefin Polymerization by Ziegler-Natta Catalysts. Vol. 12, pp. 65–86.
Casale, A. and *Porter, R. S.:* Mechanical Synthesis of Block and Graft Copolymers. Vol. 17, pp. 1–71.
Cerf, R.: La dynamique des solutions de macromolecules dans un champ de vitesses. Vol. 1, pp. 382–450.
Cesca, S., Priola, A. and *Bruzzone, M.:* Synthesis and Modification of Polymers Containing a System of Conjugated Double Bonds. Vol. 32, pp. 1–67.
Chiellini, E., Solaro R., Galli, G. and *Ledwith, A.:* Pptically Active Synthetic Polymers Containing Pendant Carbazolyl Groups. Vol. 62, pp. 143–170.
Cicchetti, O.: Mechanisms of Oxidative Photodegradation and of UV Stabilization of Polyolefins. Vol. 7, pp. 70–112.
Clark, D. T.: ESCA Applied to Polymers. Vol. 24, pp. 125–188.
Coleman, Jr., L. E. and *Meinhardt, N. A.:* Polymerization Reactions of Vinyl Ketones. Vol. 1, pp. 159–179.
Comper, W. D. and *Preston, B. N.:* Rapid Polymer Transport in Concentrated Solutions. Vol. 55, pp. 105–152.
Corner, T.: Free Radical Polymerization — The Synthesis of Graft Copolymers. Vol. 62, pp. 95–142.
Crescenzi, V.: Some Recent Studies of Polyelectrolyte Solutions. Vol. 5, pp. 358–386.
Crivello, J. V.: Cationic Polymerization — Iodonium and Sulfonium Salt Photoinitiators, Vol. 62, pp. 1–48.

Davydov, B. E. and *Krentsel, B. A.:* Progress in the Chemistry of Polyconjugated Systems. Vol. 25, pp. 1–46.
Dettenmaier, M.: Intrinsic Crazes in Polycarbonate Phenomenology and Molecular Interpretation of a New Phenomenon. Vol. 52/53, pp. 57–104.
Dobb, M. G. and *McIntyre, J. E.:* Properties and Applications of Liquid-Crystalline Main-Chain Polymers. Vol. 60/61, pp. 61–98.
Döll, W.: Optical Interference Measurements and Fracture Mechanics Analysis of Crack Tip Craze Zones. Vol. 52/53, pp. 105–168
Doi, Y. see *Keii, T.:* Vol. 73/74, pp. 201–248.
Dole, M.: Calorimetric Studies of States and Transitions in Solid High Polymers. Vol. 2, pp. 221–274.
Dorn, K., Hupfer, B., and *Ringsdorf, H.:* Polymeric Monolayers and Liposomes as Models for Biomembranes How to Bridge the Gap Between Polymer Science and Membrane Biology? Vol. 64, pp. 1–54.
Dreyfuss, P. and *Dreyfuss, M. P.:* Polytetrahydrofuran. Vol. 4, pp. 528–590.
Drobnik, J. and *Rypáček, F.:* Soluble Synthetic Polymers in Biological Systems. Vol. 57, pp. 1–50.

Dröscher, M.: Solid State Extrusion of Semicrystalline Copolymers. Vol. 47, pp. 120–138.
Drzal, L. T.: The Interphase in Epoxy Composites. Vol. 75, pp. 1–32.
Dušek, K. and *Prins, W.:* Structure and Elasticity of Non-Crystalline Polymer Networks. Vol. 6, pp. 1–102.
Duncan, R. and *Kopeček, J.:* Soluble Synthetic Polymers as Potential Drug Carriers. Vol. 57, pp. 51–101.

Eastham, A. M.: Some Aspects of the Polymerization of Cyclic Ethers. Vol. 2, pp. 18–50.
Ehrlich, P. and *Mortimer, G. A.:* Fundamentals of the Free-Radical Polymerization of Ethylene. Vol. 7, pp. 386–448.
Eisenberg, A.: Ionic Forces in Polymers. Vol. 5, pp. 59–112.
Eiss, N. S. Jr. see Yorkgitis, E. M. Vol. 72, pp. 79–110.
Elias, H.-G., Bareiss, R. und *Watterson, J. G.:* Mittelwerte des Molekulargewichts und anderer Eigenschaften. Vol. 11, pp. 111–204.
Elsner, G., Riekel, Ch. and *Zachmann, H. G.:* Synchrotron Radiation Physics. Vol. 67, pp. 1–58.
Elyashevich, G. K.: Thermodynamics and Kinetics of Orientational Crystallization of Flexible-Chain Polymers. Vol. 43, pp. 207–246.
Enkelmann, V.: Structural Aspects of the Topochemical Polymerization of Diacetylenes. Vol. 63. pp. 91–136.

Ferruti, P. and *Barbucci, R.:* Linear Amino Polymers: Synthesis, Protonation and Complex Formation. Vol. 58, pp. 55–92.
Finkelmann, H. and *Rehage, G.:* Liquid Crystal Side-Chain Polymers. Vol. 60/61, pp. 99–172.
Fischer, H.: Freie Radikale während der Polymerisation, nachgewiesen und identifiziert durch Elektronenspinresonanz. Vol. 5, pp. 463–530.
Flory, P. J.: Molecular Theory of Liquid Crystals. Vol. 59, pp. 1–36.
Ford, W. T. and *Tomoi, M.:* Polymer-Supported Phase Transfer Catalysts Reaction Mechanisms. Vol. 55, pp. 49–104.
Fradet, A. and *Maréchal, E.:* Kinetics and Mechanisms of Polyesterifications. I. Reactions of Diols with Diacids. Vol. 43, pp. 51–144.
Friedrich, K.: Crazes and Shear Bands in Semi-Crystalline Thermoplastics. Vol. 52/53, pp. 225–274
Fujita, H.: Diffusion in Polymer-Diluent Systems. Vol. 3, pp. 1–47.
Funke, W.: Über die Strukturaufklärung vernetzter Makromoleküle, insbesondere vernetzter Polyesterharze, mit chemischen Methoden. Vol. 4, pp. 157–235.

Gal'braikh, L. S. and *Rigovin, Z. A.:* Chemical Transformation of Cellulose. Vol. 14, pp. 87–130.
Galli, G. see Chiellini, E. Vol. 62, pp. 143–170.
Gallot, B. R. M.: Preparation and Study of Block Copolymers with Ordered Structures, Vol. 29, pp. 85–156.
Gandini, A.: The Behaviour of Furan Derivatives in Polymerization Reactions. Vol. 25, pp. 47–96.
Gandini, A. and *Cheradame, H.:* Cationic Polymerization. Initiation with Alkenyl Monomers. Vol. 34/35, pp. 1–289.
Geckeler, K., Pillai, V. N. R., and *Mutter, M.:* Applications of Soluble Polymeric Supports. Vol. 39, pp. 65–94.
Gerrens, H.: Kinetik der Emulsionspolymerisation. Vol. 1, pp. 234–328.
Ghiggino, K. P., Roberts, A. J. and *Phillips, D.:* Time-Resolved Fluorescence Techniques in Polymer and Biopolymer Studies. Vol. 40, pp. 69–167.
Goethals, E. J.: The Formation of Cyclic Oligomers in the Cationic Polymerization of Heterocycles. Vol. 23, pp. 103–130.
Graessley, W. W.: The Entanglement Concept in Polymer Rheology. Vol. 16, pp. 1–179.
Graessley, W. W.: Entangled Linear, Branched and Network Polymer Systems. Molecular Theories. Vol. 47, pp. 67–117.
Grebowicz, J. see Wunderlich, B. Vol. 60/61, pp. 1–60.
Greschner, G. S.: Phase Distribution Chromatography. Possibilities and Limitations. Vol. 73/74, pp. 1–62.

Hagihara, N., Sonogashira, K. and *Takahashi, S.:* Linear Polymers Containing Transition Metals in the Main Chain. Vol. 41, pp. 149–179.
Hasegawa, M.: Four-Center Photopolymerization in the Crystalline State. Vol. 42, pp. 1–49.
Hay, A. S.: Aromatic Polyethers. Vol. 4, pp. 496–527.
Hayakawa, R. and *Wada, Y.:* Piezoelectricity and Related Properties of Polymer Films. Vol. 11, pp. 1–55.
Heidemann, E. and *Roth, W.:* Synthesis and Investigation of Collagen Model Peptides. Vol. 43, pp. 145–205.
Heitz, W.: Polymeric Reagents. Polymer Design, Scope, and Limitations. Vol. 23, pp. 1–23.
Helfferich, F.: Ionenaustausch. Vol. 1, pp. 329–381.
Hendra, P. J.: Laser-Raman Spectra of Polymers. Vol. 6, pp. 151–169.
Hendrix, J.: Position Sensitive "X-ray Detectors". Vol. 67, pp. 59–98.
Henrici-Olivé, G. und *Olivé, S.:* Kettenübertragung bei der radikalischen Polymerisation. Vol. 2, pp. 496–577.
Henrici-Olivé, G. und *Olivé, S.:* Koordinative Polymerisation an löslichen Übergangsmetall-Katalysatoren. Vol. 6, pp. 421–472.
Henrici-Olivé, G. and *Olivé, S.:* Oligomerization of Ethylene with Soluble Transition-Metal Catalysts. Vol. 15, pp. 1–30.
Henrici-Olivé, G. and *Olivé, S.:* Molecular Interactions and Macroscopic Properties of Polyacrylonitrile and Model Substances. Vol. 32, pp. 123–152.
Henrici-Olivé, G. and *Olivé, S.:* The Chemistry of Carbon Fiber Formation from Polyacrylonitrile. Vol. 51, pp. 1–60.
Hermans, Jr., J., Lohr, D. and *Ferro, D.:* Treatment of the Folding and Unfolding of Protein Molecules in Solution According to a Lattic Model. Vol. 9, pp. 229–283.
Higashimura, T. and *Sawamoto, M.:* Living Polymerization and Selective Dimerization: Two Extremes of the Polymer Synthesis by Cationic Polymerization. Vol. 62, pp. 49–94.
Hoffman, A. S.: Ionizing Radiation and Gas Plasma (or Glow) Discharge Treatments for Preparation of Novel Polymeric Biomaterials. Vol. 57, pp. 141–157.
Holzmüller, W.: Molecular Mobility, Deformation and Relaxation Processes in Polymers. Vol. 26, pp. 1–62.
Hutchison, J. and *Ledwith, A.:* Photoinitiation of Vinyl Polymerization by Aromatic Carbonyl Compounds. Vol. 14, pp. 49–86.

Iizuka, E.: Properties of Liquid Crystals of Polypeptides: with Stress on the Electromagnetic Orientation. Vol. 20, pp. 79–107.
Ikada, Y.: Characterization of Graft Copolymers. Vol. 29, pp. 47–84.
Ikada, Y.: Blood-Compatible Polymers. Vol. 57, pp. 103–140.
Imanishi, Y.: Synthese, Conformation, and Reactions of Cyclic Peptides. Vol. 20, pp. 1–77.
Inagaki, H.: Polymer Separation and Characterization by Thin-Layer Chromatography. Vol. 24, pp. 189–237.
Inoue, S.: Asymmetric Reactions of Synthetic Polypeptides. Vol. 21, pp. 77–106.
Ise, N.: Polymerizations under an Electric Field. Vol. 6, pp. 347–376.
Ise, N.: The Mean Activity Coefficient of Polyelectrolytes in Aqueous Solutions and Its Related Properties. Vol. 7, pp. 536–593.
Isihara, A.: Intramolecular Statistics of a Flexible Chain Molecule. Vol. 7, pp. 449–476.
Isihara, A.: Irreversible Processes in Solutions of Chain Polymers. Vol. 5, pp. 531–567.
Isihara, A. and *Guth, E.:* Theory of Dilute Macromolecular Solutions. Vol. 5, pp. 233–260.
Iwatsuki, S.: Polymerization of Quinodimethane Compounds. Vol. 58, pp. 93–120.

Janeschitz-Kriegl, H.: Flow Birefrigence of Elastico-Viscous Polymer Systems. Vol. 6, pp. 170–318.
Jenkins, R. and *Porter, R. S.:* Uperturbed Dimensions of Stereoregular Polymers. Vol. 36, pp. 1–20.
Jenngins, B. R.: Electro-Optic Methods for Characterizing Macromolecules in Dilute Solution. Vol. 22, pp. 61–81.
Johnston, D. S.: Macrozwitterion Polymerization. Vol. 42, pp. 51–106.

Kamachi, M.: Influence of Solvent on Free Radical Polymerization of Vinyl Compounds. Vol. 38, pp. 55–87.

Kaneko, M. and *Yamada, A.:* Solar Energy Conversion by Functional Polymers. Vol. 55, pp. 1–48.

Kawabata, S. and *Kawai, H.:* Strain Energy Density Functions of Rubber Vulcanizates from Biaxial Extension. Vol. 24, pp. 89–124.

Keii, T., Doi, Y.: Synthesis of "Living" Polyolefins with Soluble Ziegler-Natta Catalysts and Application to Block Copolymerization. Vol. 73/74, pp. 201–248.

Kennedy, J. P. and *Chou, T.:* Poly(isobutylene-co-β-Pinene): A New Sulfur Vulcanizable, Ozone Resistant Elastomer by Cationic Isomerization Copolymerization. Vol. 21, pp. 1–39.

Kennedy, J. P. and *Delvaux, J. M.:* Synthesis, Characterization and Morphology of Poly(butadiene-g-Styrene). Vol. 38, pp. 141–163.

Kennedy, J. P. and *Gillham, J. K.:* Cationic Polymerization of Olefins with Alkylaluminium Initiators. Vol. 10, pp. 1–33.

Kennedy, J. P. and *Johnston, J. E.:* The Cationic Isomerization Polymerization of 3-Methyl-1-butene and 4-Methyl-1-pentene. Vol. 19, pp. 57–95.

Kennedy, J. P. and *Langer, Jr., A. W.:* Recent Advances in Cationic Polymerization. Vol. 3, pp. 508–580.

Kennedy, J. P. and *Otsu, T.:* Polymerization with Isomerization of Monomer Preceding Propagation. Vol. 7, pp. 369–385.

Kennedy, J. P. and *Rengachary, S.:* Correlation Between Cationic Model and Polymerization Reactions of Olefins. Vol. 14, pp. 1–48.

Kennedy, J. P. and *Trivedi, P. D.:* Cationic Olefin Polymerization Using Alkyl Halide — Alkylaluminium Initiator Systems. I. Reactivity Studies. II. Molecular Weight Studies. Vol. 28, pp. 83–151.

Kennedy, J. P., Chang, V. S. C. and *Guyot, A.:* Carbocationic Synthesis and Characterization of Polyolefins with Si–H and Si–Cl Head Groups. Vol. 43, pp. 1–50.

Khoklov, A. R. and *Grosberg, A. Yu.:* Statistical Theory of Polymeric Lyotropic Liquid Crystals. Vol. 41, pp. 53–97.

Kinloch, A. J.: Mechanics and Mechanisms of Fracture of Thermosetting Epoxy Polymers. Vol. 72, pp. 45–68.

Kissin, Yu. V.: Structures of Copolymers of High Olefins. Vol. 15, pp. 91–155.

Kitagawa, T. and *Miyazawa, T.:* Neutron Scattering and Normal Vibrations of Polymers. Vol. 9, pp. 335–414.

Kitamaru, R. and *Horii, F.:* NMR Approach to the Phase Structure of Linear Polyethylene. Vol. 26, pp. 139–180.

Knappe, W.: Wärmeleitung in Polymeren. Vol. 7, pp. 477–535.

Koenig, J. L.: Fourier Transforms Infrared Spectroscopy of Polymers, Vol. 54, pp. 87–154.

Koenig, J. L. see *Mertzel, E.:* Vol. 75, pp. 73–112.

Kolařik, J.: Secondary Relaxations in Glassy Polymers: Hydrophilic Polymethacrylates and Polyacrylates: Vol. 46, pp. 119–161.

Koningsveld, R.: Preparative and Analytical Aspects of Polymer Fractionation. Vol. 7.

Kovacs, A. J.: Transition vitreuse dans les polymers amorphes. Etude phénoménologique. Vol. 3, pp. 394–507.

Krässig, H. A.: Graft Co-Polymerization of Cellulose and Its Derivatives. Vol. 4, pp. 111–156.

Kramer, E. J.: Microscopic and Molecular Fundamentals of Crazing. Vol. 52/53, pp. 1–56

Kraus, G.: Reinforcement of Elastomers by Carbon Black. Vol. 8, pp. 155–237.

Kreutz, W. and *Welte, W.:* A General Theory for the Evaluation of X-Ray Diagrams of Biomembranes and Other Lamellar Systems. Vol. 30, pp. 161–225.

Krimm, S.: Infrared Spectra of High Polymers. Vol. 2, pp. 51–72.

Kuhn, W., Ramel, A., Walters, D. H., Ebner, G. and *Kuhn, H. J.:* The Production of Mechanical Energy from Different Forms of Chemical Energy with Homogeneous and Cross-Striated High Polymer Systems. Vol. 1, pp. 540–592.

Kunitake, T. and *Okahata, Y.:* Catalytic Hydrolysis by Synthetic Polymers. Vol. 20, pp. 159–221.

Kurata, M. and *Stockmayer, W. H.:* Intrinsic Viscosities and Unperturbed Dimensions of Long Chain Molecules. Vol. 3, pp. 196–312.

Ledwith, A. and *Sherrington, D. C.:* Stable Organic Cation Salts: Ion Pair Equilibria and Use in Cationic Polymerization. Vol. 19, pp. 1–56.
Ledwith, A. see Chiellini, E. Vol. 62, pp. 143–170.
Lee, C.-D. S. and *Daly, W. H.:* Mercaptan-Containing Polymers. Vol. 15, pp. 61–90.
Lindberg, J. J. and *Hortling, B.:* Cross Polarization — Magic Angle Spinning NMR Studies of Carbohydrates and Aromatic Polymers. Vol. 66, pp. 1–22.
Lipatov, Y. S.: Relaxation and Viscoelastic Properties of Heterogeneous Polymeric Compositions. Vol. 22, pp. 1–59.
Lipatov, Y. S.: The Iso-Free-Volume State and Glass Transitions in Amorphous Polymers: New Development of the Theory. Vol. 26, pp. 63–104.
Lustoň, J. and *Vašš, F.:* Anionic Copolymerization of Cyclic Ethers with Cyclic Anhydrides. Vol. 56, pp. 91–133.

Madec, J.-P. and *Maréchal, E.:* Kinetics and Mechanisms of Polyesterifications. II. Reactions of Diacids with Diepoxides. Vol. 71, pp. 153–228.
Mano, E. B. and *Coutinho, F. M. B.:* Grafting on Polyamides. Vol. 19, pp. 97–116.
Maréchal, E. see Madec, J.-P. Vol. 71, pp. 153–228.
Mark, J. E.: The Use of Model Polymer Networks to Elucidate Molecular Aspects of Rubberlike Elasticity. Vol. 44, pp. 1–26.
Mark, J. E. see Queslel, J. P. Vol. 71, pp. 229–248.
Maser, F., Bode, K., Pillai, V. N. R. and *Mutter, M.:* Conformational Studies on Model Peptides. Their Contribution to Synthetic, Structural and Functional Innovations on Proteins. Vol. 65, pp. 177–214.
McGrath, J. E. see Yorkgitis, E. M. Vol. 72, pp. 79–110.
McIntyre, J. E. see Dobb, M. G. Vol. 60/61, pp. 61–98.
Meerwall v., E., D.: Self-Diffusion in Polymer Systems. Measured with Field-Gradient Spin Echo NMR Methods, Vol. 54, pp. 1–29.
Mengoli, G.: Feasibility of Polymer Film Coating Through Electroinitiated Polymerization in Aqueous Medium. Vol. 33, pp. 1–31.
Mertzel, E., Koenig, J. L.: Application of FT-IR and NMR to Epoxy Resins. Vol. 75, pp. 73–112.
Meyerhoff, G.: Die viscosimetrische Molekulargewichtsbestimmung von Polymeren. Vol. 3, pp. 59–105.
Millich, F.: Rigid Rods and the Characterization of Polyisocyanides. Vol. 19, pp. 117–141.
Möller, M.: Cross Polarization — Magic Angle Sample Spinning NMR Studies. With Respect to the Rotational Isomeric States of Saturated Chain Molecules. Vol. 66, pp. 59–80.
Morawetz, H.: Specific Ion Binding by Polyelectrolytes. Vol. 1, pp. 1–34.
Morgan, R. J.: Structure-Property Relations of Epoxies Used as Composite Matrices. Vol. 72, pp. 1–44.
Morin, B. P., Breusova, I. P. and *Rogovin, Z. A.:* Structural and Chemical Modifications of Cellulose by Graft Copolymerization. Vol. 42, pp. 139–166.
Mulvaney, J. E., Oversberger, C. C. and *Schiller, A. M.:* Anionic Polymerization. Vol. 3, pp. 106–138.

Nakase, Y., Kurijama, I. and *Odajima, A.:* Analysis of the Fine Structure of Poly(Oxymethylene) Prepared by Radiation-Induced Polymerization in the Solid State. Vol. 65, pp. 79–134.
Neuse, E.: Aromatic Polybenzimidazoles. Syntheses, Properties, and Applications. Vol. 47, pp. 1–42.
Nicolais, L. see Apicella, A. Vol. 72, pp. 69–78.
Nuyken, O., Weidner, R.: Graft and Block Copolymers via Polymeric Azo Initiators. Vol. 73/74, pp. 145–200.

Ober, Ch. K., Jin, J.-I. and *Lenz, R. W.:* Liquid Crystal Polymers with Flexible Spacers in the Main Chain. Vol. 59, pp. 103–146.
Okubo, T. and *Ise, N.:* Synthetic Polyelectrolytes as Models of Nucleic Acids and Esterases. Vol. 25, pp. 135–181.
Osaki, K.: Viscoelastic Properties of Dilute Polymer Solutions. Vol. 12, pp. 1–64.

Oster, G. and *Nishijima, Y.:* Fluorescence Methods in Polymer Science. Vol. 3, pp. 313–331.
Otsu, T. see Sato, T. Vol. 71, pp. 41–78.
Overberger, C. G. and *Moore, J. A.:* Ladder Polymers. Vol. 7, pp. 113–150.

Packirisamy, S. see Biswas, M. Vol. 70, pp. 71–118.
Papkov, S. P.: Liquid Crystalline Order in Solutions of Rigid-Chain Polymers. Vol. 59, pp. 75–102.
Patat, F., Killmann, E. und *Schiebener, C.:* Die Absorption von Makromolekülen aus Lösung. Vol. 3, pp. 332–393.
Patterson, G. D.: Photon Correlation Spectroscopy of Bulk Polymers. Vol. 48, pp. 125–159.
Penczek, S., Kubisa, P. and *Matyjaszewski, K.:* Cationic Ring-Opening Polymerization of Heterocyclic Monomers. Vol. 37, pp. 1–149.
Penczek, S., Kubisa, P. and *Matyjaszewski, K.:* Cationic Ring-Opening Polymerization; 2. Synthetic Applications. Vol. 68/69, pp. 1–298.
Peticolas, W. L.: Inelastic Laser Light Scattering from Biological and Synthetic Polymers. Vol. 9, pp. 285–333.
Petropoulos, J. H.: Membranes with Non-Homogeneous Sorption Properties. Vol. 64, pp. 85–134.
Pino, P.: Optically Active Addition Polymers. Vol. 4, pp. 393–456.
Pitha, J.: Physiological Activities of Synthetic Analogs of Polynucleotides. Vol. 50, pp. 1–16.
Platé, N. A. and *Noak, O. V.:* A Theoretical Consideration of the Kinetics and Statistics of Reactions of Functional Groups of Macromolecules. Vol. 31, pp. 133–173.
Platé, N. A. see Shibaev, V. P. Vol. 60/61, pp. 173–252.
Plesch, P. H.: The Propagation Rate-Constants in Cationic Polymerisations. Vol. 8, pp. 137–154.
Porod, G.: Anwendung und Ergebnisse der Röntgenkleinwinkelstreuung in festen Hochpolymeren. Vol. 2, pp. 363–400.
Pospíšil, J.: Transformations of Phenolic Antioxidants and the Role of Their Products in the Long-Term Properties of Polyolefins. Vol. 36, pp. 69–133.
Postelnek, W., Coleman, L. E., and *Lovelace, A. M.:* Fluorine-Containing Polymers. I. Fluorinated Vinyl Polymers with Functional Groups, Condensation Polymers, and Styrene Polymers. Vol. 1, pp. 75–113.

Queslel, J. P. and *Mark, J. E.:* Molecular Interpretation of the Moduli of Elastomeric Polymer Networks of Know Structure. Vol. 65, pp. 135–176.
Queslel, J. P. and *Mark, J. E.:* Swelling Equilibrium Studies of Elastomeric Network Structures. Vol. 71, pp. 229–248.

Rehage, G. see Finkelmann, H. Vol. 60/61, pp. 99–172.
Rempp, P. F. and *Franta, E.:* Macromonomers: Synthesis, Characterization and Applications. Vol. 58, pp. 1–54.
Rempp, P., Herz, J., and *Borchard, W.:* Model Networks. Vol. 26, pp. 107–137.
Richards, R. W.: Small Angle Neutron Scattering from Block Copolymers. Vol. 71, pp. 1–40.
Rigbi, Z.: Reinforcement of Rubber by Carbon Black. Vol. 36, pp. 21–68.
Rogovin, Z. A. and *Gabrielyan, G. A.:* Chemical Modifications of Fibre Forming Polymers and Copolymers of Acrylonitrile. Vol. 25, pp. 97–134.
Roha, M.: Ionic Factors in Steric Control. Vol. 4, pp. 353–392.
Roha, M.: The Chemistry of Coordinate Polymerization of Dienes. Vol. 1, pp. 512–539.
Rostami, S. see Walsh, D. J. Vol. 70, pp. 119–170.
Rozenberg, B. A.: Kinetics, Thermodynamics and Mechanism of Reactions of Epoxy Oligomers with Amines. Vol. 75, pp. 113–166.

Safford, G. J. and *Naumann, A. W.:* Low Frequency Motions in Polymers as Measured by Neutron Inelastic Scattering. Vol. 5, pp. 1–27.
Sato, T. and *Otsu, T.:* Formation of Living Propagating Radicals in Microspheres and Their Use in the Synthesis of Block Copolymers. Vol. 71, pp. 41–78.
Sauer, J. A. and *Chen, C. C.:* Crazing and Fatigue Behavior in One and Two Phase Glassy Polymers. Vol. 52/53, pp. 169–224

Sawamoto, M. see *Higashimura, T.* Vol. 62, pp. 49–94.
Schmidt, R. G., Bell, J. P.: Epoxy Adhesion to Metals. Vol. 75, pp. 33–72.
Schuerch, C.: The Chemical Synthesis and Properties of Polysaccharides of Biomedical Interest. Vol. 10, pp. 173–194.
Schulz, R. C. und *Kaiser, E.:* Synthese und Eigenschaften von optisch aktiven Polymeren. Vol. 4, pp. 236–315.
Seanor, D. A.: Charge Transfer in Polymers. Vol. 4, pp. 317–352.
Semerak, S. N. and *Frank, C. W.:* Photophysics of Excimer Formation in Aryl Vinyl Polymers, Vol. 54, pp. 31–85.
Seidl, J., Malinský, J., Dušek, K. und *Heitz, W.:* Makroporöse Styrol-Divinylbenzol-Copolymere und ihre Verwendung in der Chromatographie und zur Darstellung von Ionenaustauschern. Vol. 5, pp. 113–213.
Semjonow, V.: Schmelzviskositäten hochpolymerer Stoffe. Vol. 5, pp. 387–450.
Semlyen, J. A.: Ring-Chain Equilibria and the Conformations of Polymer Chains. Vol. 21, pp. 41–75.
Sen, A.: The Copolymerization of Carbon Monoxide with Olefins. Vol. 73/74, pp. 125–144.
Sharkey, W. H.: Polymerizations Through the Carbon-Sulphur Double Bond. Vol. 17, pp. 73–103.
Shibaev, V. P. and *Platé, N. A.:* Thermotropic Liquid-Crystalline Polymers with Mesogenic Side Groups. Vol. 60/61, pp. 173–252.
Shimidzu, T.: Cooperative Actions in the Nucleophile-Containing Polymers. Vol. 23, pp. 55–102.
Shutov, F. A.: Foamed Polymers Based on Reactive Oligomers, Vol. 39, pp. 1–64.
Shutov, F. A.: Foamed Polymers. Cellular Structure and Properties. Vol. 51, pp. 155–218.
Shutov, F. A.: Syntactic Polymer Foams. Vol. 73/74, pp. 63–124.
Siesler, H. W.: Rheo-Optical Fourier-Transform Infrared Spectroscopy: Vibrational Spectra and Mechanical Properties of Polymers. Vol. 65, pp. 1–78.
Silvestri, G., Gambino, S., and *Filardo, G.:* Electrochemical Production of Initiators for Polymerization Processes. Vol. 38, pp. 27–54.
Sixl, H.: Spectroscopy of the Intermediate States of the Solid State Polymerization Reaction in Diacetylene Crystals. Vol. 63, pp. 49–90.
Slichter, W. P.: The Study of High Polymers by Nuclear Magnetic Resonance. Vol. 1, pp. 35–74.
Small, P. A.: Long-Chain Branching in Polymers. Vol. 18.
Smets, G.: Block and Graft Copolymers. Vol. 2, pp. 173–220.
Smets, G.: Photochromic Phenomena in the Solid Phase. Vol. 50, pp. 17–44.
Sohma, J. and *Sakaguchi, M.:* ESR Studies on Polymer Radicals Produced by Mechanical Destruction and Their Reactivity. Vol. 20, pp. 109–158.
Solaro, R. see *Chiellini, E.* Vol. 62, pp. 143–170.
Sotobayashi, H. und *Springer, J.:* Oligomere in verdünnten Lösungen. Vol. 6, pp. 473–548.
Sperati, C. A. and *Starkweather, Jr., H. W.:* Fluorine-Containing Polymers. II. Polytetrafluoroethylene. Vol. 2, pp. 465–495.
Spiess, H. W.: Deuteron NMR — A new Tool for Studying Chain Mobility and Orientation in Polymers. Vol. 66, pp. 23–58.
Sprung, M. M.: Recent Progress in Silicone Chemistry. I. Hydrolysis of Reactive Silane Intermediates, Vol. 2, pp. 442–464.
Stahl, E. and *Brüderle, V.:* Polymer Analysis by Thermofractography. Vol. 30, pp. 1–88.
Stannett, V. T., Koros, W. J., Paul, D. R., Lonsdale, H. K., and *Baker, R. W.:* Recent Advances in Membrane Science and Technology. Vol. 32, pp. 69–121.
Staverman, A. J.: Properties of Phantom Networks and Real Networks. Vol. 44, pp. 73–102.
Stauffer, D., Coniglio, A. and *Adam, M.:* Gelation and Critical Phenomena. Vol. 44, pp. 103–158.
Stille, J. K.: Diels-Alder Polymerization. Vol. 3, pp. 48–58.
Stolka, M. and *Pai, D.:* Polymers with Photoconductive Properties. Vol. 29, pp. 1–45.
Stuhrmann, H.: Resonance Scattering in Macromolecular Structure Research. Vol. 67, pp. 123–164.
Subramanian, R. V.: Electroinitiated Polymerization on Electrodes. Vol. 33, pp. 35–58.
Sumitomo, H. and *Hashimoto, K.:* Polyamides as Barrier Materials. Vol. 64, pp. 55–84.
Sumitomo, H. and *Okada, M.:* Ring-Opening Polymerization of Bicyclic Acetals, Oxalactone, and Oxalactam. Vol. 28, pp. 47–82.
Szegö, L.: Modified Polyethylene Terephthalate Fibers. Vol. 31, pp. 89–131.
Szwarc, M.: Termination of Anionic Polymerization. Vol. 2, pp. 275–306.

Szwarc, M.: The Kinetics and Mechanism of N-carboxy-α-amino-acid Anhydride (NCA) Polymerization to Poly-amino Acids. Vol. 4, pp. 1–65.
Szwarc, M.: Thermodynamics of Polymerization with Special Emphasis on Living Polymers. Vol. 4, pp. 457–495.
Szwarc, M.: Living Polymers and Mechanisms of Anionic Polymerization. Vol. 49, pp. 1–175.

Takahashi, A. and *Kawaguchi, M.:* The Structure of Macromolecules Adsorbed on Interfaces. Vol. 46, pp. 1–65.
Takemoto, K. and *Inaki, Y.:* Synthetic Nucleic Acid Analogs. Preparation and Interactions. Vol. 41, pp. 1–51.
Tani, H.: Stereospecific Polymerization of Aldehydes and Epoxides. Vol. 11, pp. 57–110.
Tate, B. E.: Polymerization of Itaconic Acid and Derivatives. Vol. 5, pp. 214–232.
Tazuke, S.: Photosensitized Charge Transfer Polymerization. Vol. 6, pp. 321–346.
Teramoto, A. and *Fujita, H.:* Conformation-dependent Properties of Synthetic Polypeptides in the Helix-Coil Transition Region. Vol. 18, pp. 65–149.
Theocaris, P. S.: The Mesophase and its Influence on the Mechanical Behavior of Composites. Vol. 66, pp. 149–188.
Thomas, W. M.: Mechanismus of Acrylonitrile Polymerization. Vol. 2, pp. 401–441.
Tieke, B.: Polymerization of Butadiene and Butadiyne (Diacetylene) Derivatives in Layer Structures. Vol. 71, pp. 79–152.
Tobolsky, A. V. and *DuPré, D. B.:* Macromolecular Relaxation in the Damped Torsional Oscillator and Statistical Segment Models. Vol. 6, pp. 103–127.
Tosi, C. and *Ciampelli, F.:* Applications of Infrared Spectroscopy to Ethylene-Propylene Copolymers. Vol. 12, pp. 87–130.
Tosi, C.: Sequence Distribution in Copolymers: Numerical Tables. Vol. 5, pp. 451–462.
Tran, C. see *Yorkgitis, E. M.* Vol. 72, pp. 79–110.
Tsuchida, E. and *Nishide, H.:* Polymer-Metal Complexes and Their Catalytic Activity. Vol. 24, pp. 1–87.
Tsuji, K.: ESR Study of Photodegradation of Polymers. Vol. 12, pp. 131–190.
Tsvetkov, V. and *Andreeva, L.:* Flow and Electric Birefringence in Rigid-Chain Polymer Solutions. Vol. 39, pp. 95–207.
Tuzar, Z., Kratochvil, P., and *Bohdanecký, M.:* Dilute Solution Properties of Aliphatic Polyamides. Vol. 30, pp. 117–159.

Uematsu, I. and *Uematsu, Y.:* Polypeptide Liquid Crystals. Vol. 59, pp. 37–74.

Valvassori, A. and *Sartori, G.:* Present Status of the Multicomponent Copolymerization Theory. Vol. 5, pp. 28–58.
Viovy, J. L. and *Monnerie, L.:* Fluorescence Anisotropy Technique Using Synchrotron Radiation as a Powerful Means for Studying the Orientation Correlation Functions of Polymer Chains. Vol. 67, pp. 99–122.
Voigt-Martin, I.: Use of Transmission Electron Microscopy to Obtain Quantitative Information About Polymers. Vol. 67, pp. 195–218.
Voorn, M. J.: Phase Separation in Polymer Solutions. Vol. 1, pp. 192–233.

Walsh, D. J., Rostami, S.: The Miscibility of High Polymers: The Role of Specific Interactions. Vol. 70, pp. 119–170.
Ward, I. M.: Determination of Molecular Orientation by Spectroscopic Techniques. Vol. 66, pp. 81–116.
Ward, I.M.: The Preparation, Structure and Properties of Ultra-High Modulus Flexible Polymers. Vol. 70, pp. 1–70.
Weidner, R. see *Nuyken, O.:* Vol. 73/74, pp. 145–200.
Werber, F. X.: Polymerization of Olefins on Supported Catalysts. Vol. 1, pp. 180–191.

Wichterle, O., Šebenda, J., and *Králíček, J.:* The Anionic Polymerization of Caprolactam. Vol. 2, pp. 578–595.
Wilkes, G. L.: The Measurement of Molecular Orientation in Polymeric Solids. Vol. 8, pp. 91–136.
Wilkes, G. L. see Yorkgitis, E. M. Vol. 72, pp. 79–110.
Williams, G.: Molecular Aspects of Multiple Dielectric Relaxation Processes in Solid Polymers. Vol. 33, pp. 59–92.
Williams, J. G.: Applications of Linear Fracture Mechanics. Vol. 27, pp. 67–120.
Wöhrle, D.: Polymere aus Nitrilen. Vol. 10, pp. 35–107.
Wöhrle, D.: Polymer Square Planar Metal Chelates for Science and Industry. Synthesis, Properties and Applications. Vol. 50, pp. 45–134.
Wolf, B. A.: Zur Thermodynamik der enthalpisch und der entropisch bedingten Entmischung von Polymerlösungen. Vol. 10, pp. 109–171.
Woodward, A. E. and *Sauer, J. A.:* The Dynamic Mechanical Properties of High Polymers at Low Temperatures. Vol. 1, pp. 114–158.
Wunderlich, B.: Crystallization During Polymerization. Vol. 5, pp. 568–619.
Wunderlich, B. and *Baur, H.:* Heat Capacities of Linear High Polymers. Vol. 7, pp. 151–368.
Wunderlich, B. and *Grebowicz, J.:* Thermotropic Mesophases and Mesophase Transitions of Linear, Flexible Macromolecules. Vol. 60/61, pp. 1–60.
Wrasidlo, W.: Thermal Analysis of Polymers. Vol. 13, pp. 1–99.

Yamashita, Y.: Random and Black Copolymers by Ring-Opening Polymerization. Vol. 28, pp. 1–46.
Yamazaki, N.: Electrolytically Initiated Polymerization. Vol. 6, pp. 377–400.
Yamazaki, N. and *Higashi, F.:* New Condensation Polymerizations by Means of Phosphorus Compounds. Vol. 38, pp. 1–25.
Yokoyama, Y. and *Hall, H. K.:* Ring-Opening Polymerization of Atom-Bridged and Bond-Bridged Bicyclic Ethers, Acetals and Orthoesters. Vol. 42, pp. 107–138.
Yorkgitis, E. M., Eiss, N. S. Jr., Tran, C., Wilkes, G. L. and *McGrath, J. E.:* Siloxane-Modified Epoxy Resins. Vol. 72, pp. 79–110.
Yoshida, H. and *Hayashi, K.:* Initiation Process of Radiation-induced Ionic Polymerization as Studied by Electron Spin Resonance. Vol. 6, pp. 401–420.
Young, R. N., Quirk, R. P. and *Fetters, L. J.:* Anionic Polymerizations of Non-Polar Monomers Involving Lithium. Vol. 56, pp. 1–90.
Yuki, H. and *Hatada, K.:* Stereospecific Polymerization of Alpha-Substituted Acrylic Acid Esters. Vol. 31, pp. 1–45.

Zachmann, H. G.: Das Kristallisations- und Schmelzverhalten hochpolymerer Stoffe. Vol. 3, pp. 581–687.
Zaikov, G. E. see Aseeva, R. M. Vol. 70, pp. 171–230.
Zakharov, V. A., Bukatov, G. D., and *Yermakov, Y. I.:* On the Mechanism of Olifin Polymerization by Ziegler-Natta Catalysts. Vol. 51, pp. 61–100.
Zambelli, A. and *Tosi, C.:* Stereochemistry of Propylene Polymerization. Vol. 15, pp. 31–60.
Zucchini, U. and *Cecchin, G.:* Control of Molecular-Weight Distribution in Polyolefins Synthesized with Ziegler-Natta Catalytic Systems. Vol. 51, pp. 101–154.

Subject Index

Absorption of moisture 27
Acidic curing agents 5
Active propagating site 151
Adhesion durability 50–59, 65
–, mechanical aspects 53–56
– mechanisms 41, 42
– –, chemical bonding 41
– –, dispersion forces 42
– –, electromagnetic interactions 41
– –, mechanical aspects 41
–, work of 45
– strength 40
– –, dry conditions 42
– –, reversibility 44
– –, wet conditions 43, 44
Adhesive bonds 39
– –, durability 39
– –, water stability 39
Adhesives 43
Adsorption 14
Alcoholysis, reactions 146
Alkali metal cations, electrophilic assistance 153
Aluminium 10
– alloy compositions 10
Amides 5
Amine adsorption 14
– mixtures 158
Amines 5, 115, 116, 146, 149
–, accelerated polymerization 159
–, as electrophilic reagents 118
–, as nucleophilic reagents 118
–, initiation 150
Aminofunctional silanes 15
Amino groups, reactivities 132, 133
Aminolysis, reaction 146
Anionic polymerisation 146
Anisotropy, local 20
– patterns 101
Aramid fibers 10
– –, surface chemistry 10
Associates, auto- 128, 129
–, cyclic 123, 124
–, hetero- 128
–, self- 124
Association, intermolecular 135
Applications 40
ATR 77
Auger electron spectrometry (AES) 64

– surface spectroscopy 10
– electron spectroscopy (AES) 53
Autocatalysis 135
Autoacceleration reactions 139
Autoassociates 128, 129
Autocomplexes 120
Autoinhibition 135
– effect 121, 133

Basic curing agents 5
Bimolecular kinetics 155
Boron trifluoride 5
Branching process theory 143
Brittle interphase 26
Brittleness 16

Carbon fibers, polyacrylonitrile precursor 21
– fiber, surface 9
Carboxylic acid anhydrides 5
Catalysis, bifunctional 119
Cation-exchange materials 59
Chain propagation effect 154
– termination reactions 156
– transfer to alcohol 156
– – to the counterion 157
Chemical resistance 15
– shifts, anisotropic 105
– –, isotropic 107
Circumferential (hoop)stress 17
Coating thickness 48, 49
Cohesive strength 43
Composition analysis 86
Conne advantage 76
Corrosion 47, 61
– inhibitors 51, 52, 59, 60
– prevention 57, 58, 59, 60
– reactions 47, 56, 59
Compensation mechanism 127, 129
Competitive adsorption 14
Composite compressive strength 18
– properties 17
Compositional differences 15
Compressive strength, composite 18
Concentration gradient 16
Coupling agents 15, 50, 51, 52, 53, 92
– –, hydrophilicity 53
Crack tip 23
Crosslinking agent purity 6
Cross-polarization 94

Crystal polymers, amorphous 107
Curing agents 36
– –, acidic 5
– –, basic 5
– –, vapor pressure 5
Curing kinetics 91
– under front propagation conditions 140
– – under nonisothermal conditions 139
Cycles, ineffective 143, 144
Cyclization, intermolecular 146
– reactions, ineffective 142

Degradation of epoxy resins 92
Dehydration 28
Desmutting 40
Desorption of moisture 27
Difference spectrum analysis 87
Diffusion 44, 57, 59
– control 135, 136
Difunctional epoxy resin 4
Dipolar decoupling 93
Dispersive IR 74
Donor-acceptor complexes 118, 120, 122, 160
– – interactions 119, 120, 128, 129, 130
DRIFT 78
Dry conditions 40
Dušek theory, cascade processes 110

Elasticity, equilibrium modules 131
Electric conductivity 151
Electrophilic assistance 118, 141, 155
– –, alkali metal cations 153
β-Elimination reaction 156
Energy dispersive x-ray spectrometry (EDX) 64
Environmental resistance of epoxy composites 27
Epoxide-alcohol reaction 145
– equivalent weight (EEW) measurements 103
Epoxy-amine interactions, thermochemistry 125
– – networks, computer-assisted modelling 138
– chemistry 41
– network, properties 36
– resins, degradation 92
– –, water absorption 92
– ring opening, abnormal 156
– – –, direction of 154
– – –, normal 156

Fabrication 12
Factor analysis 88, 91
Failure, locus 42, 43, 44, 46, 61, 62, 64
Fellget advantage 75
Fiber composite, reinforced 3
– deformation 24

– fracture 24
– -matrix debonding 23
– – interphase 3
– pull-out 25
Fibers, untreated 21
Fibril orientation 9
Fillers 60, 61
Finish 15
Flexibilizers 61
Fluorinated epoxy resins 57
Forest Products Laboratories (FPL) process 54
Formulations 6
FPL 55, 56
Fracture properties 6
– toughness 25
–, work 24
Free energy 16
– ion reactivity 154
Frequency advantage 76
Front propagation conditions, curing 140
FT, fast 75
FT-IR 74

Gas-phase desorption 12
Gelation conversion, critical 131
Gel point, critical 132
Glass fibers 10
– transition temperature 136
Grant-Cheney model, steric hindrance 109

Heteroassociates 128
Heterogeneity 6, 7
High-temperature transformation 160
Hofmann intramolecular splitting reaction of chain transfer 152
Hoop stress 17
Hydration inhibitor 55, 56
Hydrophilic surface 11
Hydroxyl groups 11

Ineffective cycles 143, 144
Infrared (IR) 74
–, dispersive 74
–, FT 74
Inhibition 159
Inhomogeneities 139
Inorganic bases 5
Interfacial fracture, locus 21
– failure, mode 22
– region, pH 59
– shear strength 21
– – stress 20
– separation 14
– stress 19
– zone 23
Interlocking, mechanical 13, 40, 41, 42, 54, 55

Subject Index

Intermolecular association 135
— complexes 122
— cyclisation reactions 146
Internal stress 48, 49, 50, 60, 61
Interphase 17
—, brittle 26
— region 4, 15
Intramolecular complexes 122
Ion scattering spectrometry (ISS) 62
— pair reactivity 155
IRS 77

Jacquinot advantage 75

Kinetic parameters 131, 141, 142
— —, effective 130
Kinetics, bimolecular 155
—, trimolecular 155
Kubelka-Munk theory 78

Least squares analysis 87
Lewis acids 5
— bases 5
Line broadening, inhomogeneous 107
Loading, longitudinal 18
Locus of failure 42, 43, 44, 46, 61, 62, 64

MAS 93
— high power decoupling 105
Mechanical interlocking 40, 41, 42, 54, 55
— properties 6
— strength 15
Mercaptoester coupling agents 51, 52
Metal oxides 37, 39, 54, 65
— —, catalytic activity 59
— —, electrical conductivity 58, 59
— —, hydration 37, 46, 55, 56
— pretreatments 38, 39, 40, 54
— —, acid etching 39
— —, alkaline hydrogen peroxide etch 40
— —, hydrofluoric acid etch 40
— —, mechanical abraison 39
— —, sulfuric acid etch 40
— surfaces 36, 65
Michelson interferometer 74
Micromechanical analysis 17
— models 20
Microtoming 21
Microtopography 16
MIR 77
Mixing 7
Mixtures, rule of 17
Model compounds 115, 122
Modulus, transverse 20
Moisture exposure 28
Molecular dynamics 103
— motion 98, 100, 106, 108
— weight 5
Monolithic character 140
Monomolecular layer 15
Morphological differences 15
Multiplex advantage 75

Native defect 21
Network formation, kinetics 130
— heterogeneity 7
— termination probability parameters 143
NMR, Al-27 94, 96
—, C-13 94, 95
—, proton 94
—, Si-29 94, 95
—, solid state C-13 100
—, solid state proton 97
—, solution multinuclear 94
Noncatalytic interaction 142
— reaction 116
Nonisothermal conditions, curing 139
Nonreactive complexes 134
Nucleophilic addition 116
— attack 141

Order of reactions 117
Organic acids, basic 5
Oxide morphology 10
Oxygen permeation 58

PAA 55, 56
Peel strength 48, 49, 52
Phase transition 126
Phenols 5
Phosphoric acid anodizing (PAA) process 54
Photoacoustic technique 78
Physisorbed species 12
— water 10
Plasticization 44, 61
— of matrix 27
Polyacrylonitrile precursor carbon fibers 21
Polycondensation 159
Polymerization rate 151
Polymer structure 148
Polyphenylglycidyl ether,
  molecular-mass distribution 149
Poly(p-phenylene terephthalamide) 10
Pores 13
Porosity 16
Primer 15
Propagating site, active 151
— effect, chain 154
Pull-out length 24
Push-pull mechanism 118

Rate constant, trimolecular 117
— constants, trimolecular 127
— —, quasi-bimolecular 127, 130

Reaction enthalpies 126
— thermodynamics 119
Reagent structure and reactivity 140
Reinforced fiber composite 3
Relaxation studies 101
— times (T_1, T_{19}, T_2) 97
Rule of mixtures 17

Scanning electron microscopy (SEM) 46, 64
— transmission electron microscope 54
Secondary ion mass spectrometry (SIMS) 62, 63
Segmental motion 98, 100
Self-associates 124
Shear 17
Shear failure, basal plane 22
— strength, composite short beam 22
— —, interfacial 21
— stress 17
— —, interfacial 20
Shrinkage 48, 49, 60
Side reactions 144
Signal-to-noise ratio 75
Silane coupling agents 50, 51
Silanes, aminofunctional 15
Solubility 5
Solubilization 12
Solvent degreasing 38
— effects 144
— evaporation rate 48, 49
Spectral resolution 102
Spin-lattice effects 102
— — fluctuation 106
Spin-spin effects 102
— — fluctuation 106
Sputtering 63
Steel 10
Strain energy 127
Strength-to-weight ratio 16
Stress concentration factor 19
Structure-property relationship 5, 6
Substitution effect 130, 132, 133
Surface area 13
— —, interfacial 13
— chemical groups 21
— chemistry 10
— contamination 38
— energies 44
— free energies 16

—, hydrophilic 11
—, metal 36
— of glass, hydrated 11
—, solid reinforcement 16
— spectroscopy 9
—, titanium oxide 14
— topography, mechanical effect 13
— -treated fibers 21

Tautomeric transformation 157
Tensile load 17
— modulus, longitudinal 17
Termination reactions, chain 156
Thermal degradation, reactions 146
— expansion coefficients 48, 49, 60, 61
Thermodynamics, parameters 121, 123, 124
—, reaction 119
Thermo-oxidative degradation, reactions 146
Titanium 10
— oxide surfaces 14
Topographical nature 14
Topological reaction limit 137, 138
Transepoxidation 145
Transformation, high-temperature 160
Transition state, cyclic 119, 129
— —, trimolecular 116, 150, 152, 154
Transmission measurements 76
Transverse modulus 20
Trimolecular kinetics 155
— mechanism 160
— rate constant 117
— transition state 116, 150, 152, 154

Untreated fibers 11, 21

Voids 12

Water absorption of epoxy resins 92
— permeation 57
Weak boundary layer (WBL) 45, 61, 62
Wettability 16
Wetting 38
—, thermodynamic 16

X-ray diffraction 46
X-ray photoelectron spectrometry (XPS) 53, 64
X-ray photoelectron spectroscopy (XPS) 55

Zwitter-ion mechanism 152